Modeling Uncertainty in
the Earth Sciences

Modeling Uncertainty in the Earth Sciences

Jef Caers

WILEY-BLACKWELL

A John Wiley & Sons, Ltd., Publication

This edition first published 2011
© 2011 by John Wiley & Sons Ltd

Wiley-Blackwell is an imprint of John Wiley & Sons, formed by the merger of Wiley's global Scientific, Technical and Medical business with Blackwell Publishing.

Registered Office
John Wiley & Sons Ltd, The Atrium, Southern Gate, Chichester, West Sussex, PO19 8SQ, UK

Editorial Offices
The Atrium, Southern Gate, Chichester, West Sussex, PO19 8SQ, UK
9600 Garsington Road, Oxford, OX4 2DQ, UK
111 River Street, Hoboken, NJ 07030-5774, USA

For details of our global editorial offices, for customer services and for information about how to apply for permission to reuse the copyright material in this book please see our website at www.wiley.com/wiley-blackwell

The right of the author to be identified as the author of this work has been asserted in accordance with the UK Copyright, Designs and Patents Act 1988.

Library of Congress Cataloguing-in-Publication Data

Caers, Jef.
    Modeling uncertainty in the earth sciences / Jef Caers.
        p. cm.
    Includes index.
    ISBN 978-1-119-99263-9 (cloth) – ISBN 978-1-119-99262-2 (pbk.)
    1. Geology–Mathematical models.   2. Earth sciences–Statistical methods.   3. Three-dimensional imaging in geology.   4. Uncertainty.   I. Title.
    QE33.2.M3C34 2012
    551.01′5195–dc22

                                                                                2011008403

A catalogue record for this book is available from the British Library.

Typeset in 10.5/13pt Times by Aptara Inc., New Delhi, India

First Impression   2011

Contents

Preface

26 June 2010: CNN headlines

Tropical storm plus oil slick equal uncertainty

BP DeepHorizon spill

Decision question: "Will BP evacuate the clean-up crew knowing that evacuation requires at least three days, with the consequence of more oil spilling in the Gulf from the deep-water well, or, will BP leave the crew, possibly exposing them to tropical storm Alex, which may or may not become a hurricane?" A simple question: what is the best decision in this case?

Whether Earth Science modeling is performed on a local, regional or global scale, for scientific or engineering purposes, uncertainty is inherently present due to lack of data and lack of understanding of the underlying phenomena and processes taking place. This book highlights the various issues, techniques and practical modeling tools available for modeling uncertainty of complex Earth systems, as well as the impact it has on practical geo-engineering decision problems.

Modeling has become a standard tool in the Earth Sciences. Atmospheric scientists build climate models, seismologists build models of the deep Earth's structure, and hydrogeologists build models of aquifers. Many books and papers have been written on modeling, spread over many subdisciplines of mathematics and the Earth Sciences. Often, one or at most a few models are built to test certain hypothesis and assumptions, to validate or test certain engineering actions taken in the real world, or to attempt to describe physical processes as realistic as possible. The issue of uncertainty (historic, present or future) is often mentioned, but more as a side note; it is still rarely used for quantitative and predictive purposes. Very few books have uncertainty in Earth Sciences modeling as a primary topic; to date, no book to my knowledge discusses this at the level an undergraduate student in the Earth Sciences can actually comprehend and master. Professionals that are not academics often get lost in the myriad of technical details, limitations and assumptions of models of uncertainty in highly technical journal publications or books.

Therefore, in 2009, I decided to teach an entirely new class at Stanford University termed "Modeling Uncertainty in the Earth Sciences," as part of the curriculum for Earth Science senior undergraduate and first year graduate students (geology, geophysics and reservoir engineers) as well as related fields (such as civil and environmental engineering and Earth systems studies). The focus of this class is not to build a single model of the Earth or of its physical processes for whatever purpose and then "add on" something related to uncertainty, but to build directly a model of uncertainty for practical decision purposes. The idea is not to start from a single estimate of a certain phenomenon and then "jiggle" the numbers a bit to get some confidence statement about that estimate. The idea is to have students think in terms of uncertainty directly, not in terms of a single climate, seismological or hydrological model or any single guess, from the beginning. The quest for a new syllabus was on.

In many discussions I had with various colleagues from various disciplines in the Earth Sciences, as well as from my decade-long experience as Director of the Stanford Center for Reservoir Forecasting, I had come to the conclusion that any modeling of uncertainty is only relevant if made dependent on the particular decision question or practical application for which such modeling is called for. This, I understand, is a rather strong statement. I strongly believe there is no "value" (certainly not in dollar terms) in spending time or resources in building models of uncertainty without focusing on what impact this uncertainty will have on the decision question at hand: do we change climate-related policies? Do we tax CO_2? Do we clean a contaminated site? Where do we drill the next well? and so on.

Let's consider this more closely: if uncertainty on some phenomenon would be "infinite", that is, everything imaginable is possible, but that uncertainty has no impact on a certain decision question posed, then why bother building any model of uncertainty in the first place, it would be a waste of time and resources! While this is an extreme example, any model approach that first builds a model of uncertainty about an Earth phenomenon and then only considers the decision question is likely to be highly inefficient and possibly also ineffective. It should be stressed that there is a clear difference between building a model of the Earth and building a model of uncertainty of the Earth. For example, building a single model of the inner Earth from earthquake data has value in terms of increasing our knowledge about the planet we live on and getting a better insight into how our planet has evolved over geological time, or will evolve in the short and long term. A model of uncertainty would require the seismologist to consider all possibilities or scenarios of the Earth structure, possibly to its finest detail, which may yield a large set of possibilities because the earthquake data cannot resolve meter or kilometer-scale details at large depths. Constructing all these possibilities is too difficult given the large computation times involved in even getting a single model. However, should the focus be on how a seismological study can determine future ground motion in a particular region and the impact on building structures, then many prior geological scenarios or subsurface possibilities may not need to be considered. This would make the task of building a model of uncertainty efficient computationally and effective in terms of the application envisioned. Knowing what matters is therefore critical to building models of uncertainty and an important topic in this book.

Thinking about uncertainty correctly or at least in a consistent fashion is tricky. This has been my experience with students and advanced researchers alike. In fact, the matter of uncertainty quantification borders the intersection of science and philosophy. Since uncertainty is related to "lack of knowledge" about what is being modeled, the immediate rather philosophical question of "what is knowledge?" arises. Even with a large amount of data, our knowledge about the universe is, by definition, limited because we are limited human beings who can only observe that which we are able to observe; we can only comprehend that which we are able to comprehend. Our "knowledge" is in constant evolution: just consider Newtonian physics, which was considered a certainty until Einstein discovered relativity resulting in the collapse of traditional mathematics and physics at that time. While this may seem a rather esoteric discussion, it does have practical consequence on how we think about uncertainty and how we approach uncertainty, even for daily practical situations. Often, uncertainty is modeled by including all those possibilities that cannot be excluded from the observations we have. I would call this the "inclusion" approach to modeling uncertainty: a list or set of alternative events or outcomes that are consistent with the information available is compiled. That list/set is a perfectly valid model of uncertainty. In this book, however, I will often argue for an "exclusion" approach to thinking about uncertainty, namely to start from all possibilities that can be imagined and then exclude those possibilities that can be rejected by any information available to us. Although the inclusion and exclusion approaches may lead to the same quantification of uncertainty, it is more likely that the exclusion approach will provide a more realistic statement of uncertainty in practice. It is a more conservative approach, for it is typical human behavior to tend to agree on including less than the remainder of possibilities after exclusion. In a group of peers we tend to agree quicker on what to include, but tend to disagree on what to exclude. In the exclusion approach one focuses primordially on all imaginable possibilities, without being too much biased from the beginning by information, data or other experts. In this way we tend to end up with having less (unpleasant) surprises ultimately. Nevertheless, at the same time, we need to recognize that both approaches are limited by the set of solutions that can be imagined, and hence by our own human knowledge of the universe, no matter what part of the universe (earth or atmosphere, for example) is being studied.

My personal practical experience with modeling uncertainty lies in the subsurface arena. The illustration example and case studies in this book contain a heavy bias towards this area. It is a difficult area for modeling uncertainty, since the subsurface is complex, the data are sparse or at best indirect, a medium exists that can be porous and/or fractured. Many applications of modeling uncertainty in the subsurface are very practical in nature and relevant to society: the exploration and extraction of natural resources, including groundwater; the storage of nuclear material and gasses such as natural gas or carbon dioxide to give a few examples. Nevertheless, this book need not be read as a manual for modeling uncertainty in the subsurface; rather, I see modeling of the subsurface as an example case study as well as illustration for modeling uncertainty in many applications with similar characteristics: complex medium, complex physics and chemistry, highly computationally complex, multidisciplinary and, most importantly, subjective in nature, but requiring a consistent repeatable approach that can be understood and communicated among the various fields of science involved. Many of the tools, workflows and methodologies presented in this book could apply to other modeling areas that have elements in common with subsurface modeling: the modeling of topology and geometry of surfaces and the modeling of spatial variation of properties (whether discrete or continuous), the assessment of response functions and physical simulation models, such as provided through physical laws. As such, the main focus of application of this book is in the area of "geo-engineering". Nevertheless, many of the modeling tools can be used for domains such as understanding fault geometries, sedimentary systems, carbonate growth systems, ecosystems, environmental sciences, seismology, soil sciences and so on.

The main aim of this book is therefore twofold: to provide an accessible, introductory overview of modeling uncertainty for the senior undergraduate or first year graduate student with interest in Earth Sciences, Environmental Sciences or Mineral and Energy Resources, and to provide a primer reading for professionals interested in the practical aspects of modeling uncertainty. As a primer, I will provide a broad rather than deep overview. The book is therefore not meant to provide an exhaustive list of all available tools for modeling uncertainty. Such book would be encyclopedic in nature and would distract the student and the first reader from the main message and most critical issues. Conceptual thinking is emphasized over theoretical understanding or encyclopedic knowledge.

Many theoretical details of the inner workings of certain methodologies are left for other, more specialized books. In colleges or universities one is used to emphasizing learning on *how* things work exactly (for example, how to solve a matrix with Gaussian elimination); as a result, often, *why* a certain tool is applied to solve a certain problem in practice is lost in the myriad of technical details and theoretical underpinnings. The aim, therefore, is to provide an overview of modeling uncertainty, not some limited aspect of it in great detail, and to understand *what* is done, why it is done that way and not necessarily *how* exactly it works (similarly, one needs to know about Gaussian elimination and what this does, but one doesn't need to remember exactly how it works unless one is looking to improve its performance). A professional will rarely have time to know exactly the inner working of all modeling techniques or rarely be involved in the detailed development of these methods. This is a book for the user, the designer of solutions to engineering problems, to create an intelligence of understanding around such design; the book is not for the advanced developer, the person who needs to design or further enhance a particular limited component in the larger workflow of solving issues related to uncertainty.

Therefore, in summary: what this book does not provide:

- An encyclopedic overview of modeling uncertainty.

- A textbook with exercises.

- A detailed mathematical manifest explaining the inner workings of each technique.

- A cook-book with recipes on how to build models of uncertainty.

- Exhaustive reference lists on every relevant paper in this area.

What this book does attempt to provide:

- A personal view on decision-driven uncertainty by the author.

- An intuitive, conceptual and illustrative overview on this important topic that cuts through the mathematical forest with the aim of illuminating the essential philosophies and components in such modeling.

- Methods, workflows and techniques that have withstood the test in the real world and are implemented in high quality commercial or open source software.

- A focus on the subsurface but with a qualification in various sections towards other applications.

- Some further suggest reading, mostly at the same level of this book.

- Teaching materials, such as slides in PDF, homework, software, and data, as well as additional material, are provided on http://uncertaintyES.stanford.edu

Acknowledgements

Many people have contributed to this book: through discussion, by providing ideas, supplying figures and other materials. First and foremost, I want to thank the students of Energy 160/260, "Modeling Uncertainty in the Earth Sciences". A classroom setting at Stanford University is the best open forum any Author could wish for. Their comments and remarks, their critical thinking has given me much insight into what is important, what strikes a chord but also where potential pitfalls in understanding lie. I would also like to thank Gregoire Mariethoz and Kiran Pande for their reviews. Reidar Bratvold provided me an early version of his book on "Making good decisions" and the discussion we had were most insightful to writing chapter 4. I thank my co-author of chapter 8, Guillaume Caumon for his contribution to structural modeling and uncertainty. Many elements of that chapter were done with the help of Nicolas Cherpeau and using the gOcad (Skua) software supplied by Paradigm. Kwangwon Park and Celine Scheidt were invaluable in writing chapters 9 and 10 on the distance-based uncertainty techniques. In that regard, I am also thankful to Schlumberger for supplying to Petrel/Ocean software that was used in some of the case studies. Esben Auken supplied useful comments to the introductory case study in Chapter 1. Mehrdad Honarkhah helped me in constructing the case example of Chapter 12 that was used as one of the projects in my course. As a teaching assistant, he went above and beyond to call of duty to make the first version of this course successful. I would also like to thank Alexandre Boucher for the use and development of the S-GEMS software as well as the supplying the figure for the artwork on the cover. Sebastien Strebelle, Tapan Mukerji, Flemming Jørgensen, Tao Sun, Holly Michael, Wikimedia and NOAA supplied essential figure materials in the book. I would like to thank the member companies of the SCRF consortium (Stanford Center for Reservoir Forecasting) for their financial support. I also would like to thank my best friends and colleagues, Margot Gerritsen and Steve Gorelick for their enthusiasm and support, in so many other things than just the writing. Finally, I want to thank Wiley-Blackwell, and in particular Izzy Canning, for giving me the opportunity to publish this work and for making the experience a smooth one!

1

Introduction

1.1 Example Application

1.1.1 Description

To illustrate the need for modeling uncertainty and the concepts, as well as tools, covered in this book, we start off with a virtual case study. "Virtual" meaning that the study concerns an actual situation in an actual area of the world; however, the data, geological studies and, most importantly, the practical outcomes of this example should not be taken as "truth," which is understandably so after reading the application case.

Much of the world's drinking water is supplied from groundwater sources. Over the past several decades, many aquifers have been compromised by surface-borne contaminants due to urban growth and farming activities. Further contamination will continue to be a threat until critical surface recharge locations are zoned as groundwater protection areas. This can only be successfully achieved if the hydraulically complex connections between the contaminant sources at the surface and the underlying aquifers are understood.

Denmark is one example of this type of scenario. Since 1999, in an effort to identify crucial recharge zones (zones where water enters the groundwater system to replenish the system), extensive geophysical data sets were collected over the Danish countryside – the areas designated as particularly valuable due to their high rate of water extraction. The data were collected with the intention of making more informed decisions regarding the designation of recharge protection zones. The magnitude of these decisions is considerable, as it could involve the relocation of farms, industry, city development and waterworks together with related large compensations. Consequently, incorrectly identifying a vulnerable area can lead to a costly error. In fact, the Danish Government set out a 10-point program (Figure 1.1) that sets certain objectives and formulates certain desired preferences, some of which may be in conflict with keeping the farming industry alive and ensuring economic health next to ecological health for this area.

The subsurface in Denmark consists of so-called buried valleys, which are considered the informal term for Pleistocene (Quaternary) subglacial channels. They have also been

Modeling Uncertainty in the Earth Sciences, First Edition. Jef Caers.
© 2011 John Wiley & Sons, Ltd. Published 2011 by John Wiley & Sons, Ltd.

Danish Government's 10-point program (1994)
Pesticides dangerous to health and environment shall be removed from the market
Pesticide tax – the consumption of pesticides shall be halved
Nitrate pollution shall be halved before 2000
Organic farming shall be encouraged
Protection of areas of special interest for drinking water
New Soil Contamination Act – waste deposits shall be cleaned up
Increased afforestation and restoration of nature to protect groundwater
Strengthening of the EU achievements
Increased control of groundwater and drinking water quality
Dialogue with the farmers and their organisations

Source: http://www.geus.dk/program-areas/water/denmark/case_groundwaterprotection_print.pdf

Figure 1.1   Objectives of the Danish Government.

described as the result of waxing and waning of Pleistocene ice sheets. The primary method by which these valleys are formed is subglacial meltwater erosion under the ice or in front of the ice margin. Thus, the valley formation is directly related to the morphology and erodability of the geological strata. The secondary method is through direct glacial erosion by ice sheets.

Several of the processes that created and filled buried valleys are important for understanding the complexity of the Danish aquifer systems and their vulnerability to surface-borne pollutants. In Denmark, the superposition of three different generations of glaciations has been observed. Thus, multigeneration glacial valleys cross-cut each other and can also appear to abruptly end (as seen in Figure 1.2). The existence and location of these glacial valleys can be thought of as the primary level of Denmark's aquifer system structure. If largely filled with sand, the buried valley has potential for being a high volume aquifer (reservoir). However, these buried valleys can be "re-used," as revealed by the observed cut-and-fill structures. This describes the secondary level of uncertainty of heterogeneity in Danish aquifer systems.

Most cut-and-fill structures are narrower than the overall buried valley, but in some places very wide structures that span the entire valley width can be seen. The complex internal structure can be observed in seismic surveys, electromagnetic surveys and occasionally in borehole data.

—Sandersen and Jorgensen (2006)

Figure 1.2 shows a few different possible internal heterogeneities and varying extent of overlying strata, which deems the valley as actually "buried."

Due to the generally complex internal structure of the valleys, potentially protective clay layers above the aquifers are likely to be discontinuous. The aquifers inside the valley will thus have a varying degree of natural protection. Even if laterally extensive clay layers are present, the protective effect will only have local importance if the surrounding

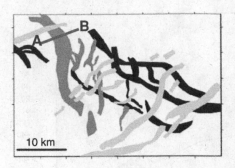
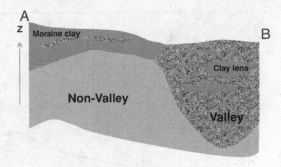

Figure 1.2 Geological interpretation of subsurface glacial channels cross-cutting each other (left). Conceptual view of the inner structure of the glacial channels (right).

sediments are sand-dominated. The valleys may therefore create short-circuits between the aquifers in the valley and the aquifers in the surrounding strata.

1.1.2 3D Modeling

In this case study, the incompleteness of the information about the subsurface strata makes making specific decisions such as relocating farms difficult. A geologist may be tempted to study in great detail the process by which these glacial valleys were created and come up with a (deterministic) description of these systems based on such understanding, possibly a computer program to simulate the process that created these systems according the physical understanding of what is understood to occur. However, such description alone will fall short in addressing the uncertainty issue that has considerable impact on the decisions made. Indeed, even if full insight into the glaciation process exists (a considerable assumption), then that would not necessarily provide a deterministic rendering of the exact location of these valleys, let alone the detailed spatial distribution of the lithologies (shale, sand, gravel, clay) inside such valleys. This does not mean that the study of the geological processes is useless. On the contrary, such study provides additional information about the possible spatial variation of such channels next to the data gathered (drilling, geophysical surveys). Therefore, additional tools are needed that allow the building of a model of the subsurface glaciations as well as quantifying the uncertainty about the spatial distribution of valley/non-valley and the various lithologies within a valley. Such a model would ideally include the physical understanding as well as reflecting the lack of knowledge, either through limited data or limited geological understanding.

Data play a crucial role in building models and constraining any model of uncertainty, whether simple or complex. In the Danish case, two types of data are present: data obtained through drilling and data obtained through a geophysical method termed time-domain electromagnetic surveys (TEM surveys). Figure 1.3 shows the interpretation of the thickness of the valleys from such surveys, which are basically a collection of 1D (vertical) soundings. The data collected are typical of many Earth modeling situations: some detailed small scale information is gathered through sampling (in this case drilling a well) and some larger scale indirect measurement(s) collected either through geophysical

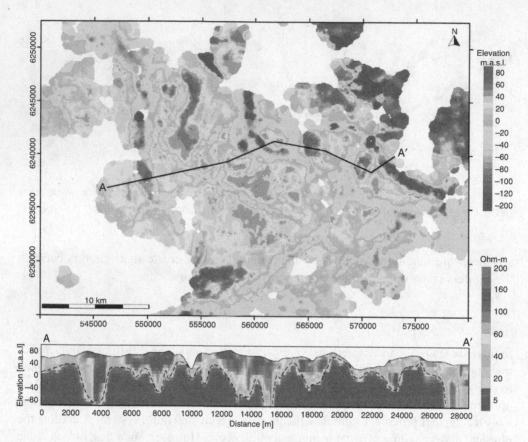

Figure 1.3 Thickness of the valley complex as processed and interpreted from TEM data. Thicker strata reflect the existence of valleys (with permission from Elsevier Science).

or remote sensing methodologies. In the Danish study, the TEM data provide a reasonably good constraint on the location of the valleys but do not inform the internal valley structure (small scale variation), while the drilling data provide the exact opposite.

1.2 Modeling Uncertainty

From this case study of modeling the subsurface, several elements in modeling uncertainty that are typical to many similar applications can be identified:

l **Decision making**: modeling uncertainty is not a goal on its own, it is usually needed because a particular decision question is raised. In fact, this decision question is usually framed in a larger context, such as done by the 10-point program, specifying objectives and preferences. Two example decisions are in this case: (1) in which areas do we relocate pollution sources and (2) do we consider taking more geophysical data to narrow the uncertainty on locating vulnerable areas, hence increasing the probability of a good decision? This latter question is termed a "Value of Information" question. Clearly, we need to make decisions without perfect information. These narrower decision questions should not be considered as independent of the larger objective outlined in Figure 1.1.

2 **Importance of the geological setting**: a critical parameter influencing the decision is the heterogeneity of the subsurface medium (fluids and soils/rocks). Rarely do we have perfect information to deterministically model the geological variability of the subsurface. Hence there is a need to model all aspects of uncertainty as related to the subsurface heterogeneity. While Figures 1.2 and 1.3 may provide one such interpretation of the system, often many alternative and competing interpretations are formed.

3 **Data**: several sources of data are available to constrain the models of uncertainty built. These data sources can be very diverse, from wells (driller's logs, well-log, cores, etc.) to geophysical (TEM data in the Danish case) or remote sensing measurements. Tying all this data into a single model of uncertainty without making too many assumptions about the relationships between various data sources is challenging.

From this case study, it is clear that some of the tools for modeling random phenomena through traditional probability models are too rigid to handle all these complexities. The nature of modeling uncertainty in the Earth Science has various challenge and issues that need to be addressed.

1 Modeling uncertainty is often application tailored. If the application changes then the type of modeling and the approach to modeling uncertainty will be different, hence the model of uncertainty will be different. Building a model of uncertainty that includes all possible aspects of what is uncertain is too difficult and often not needed in the first place. Modeling uncertainty for the sake of uncertainty is basically irrelevant as well as an impossible task. For example, if one is looking to quantify the global reserves of an oil reservoir, then the focus should be on the structural model and global parameters such as net-to-gross, while if the question is about drilling the next well, than the analysis should focus on local reservoir heterogeneity and connectivity of flow units.

2 Several sources of uncertainty exist for this case study:

   a Uncertainty related to the measurement errors and processing of the raw measurements.

   b Uncertainty related to the fact that processed data can be interpreted in many ways and, in fact, that data interpretation and processing require a model on their own.

   c Uncertainty related to the type of geological setting used, which is interpreted from data or based on physical models which themselves are uncertain.

   d Spatial uncertainty: even if data were perfectly measured, they are still sparse with respect to the resolution at which we want to build models. This means that various models with different spatial distributions of properties or layering structures can be generated matching equally well the same data.

  e Response uncertainty: this includes uncertainty related to how geological uncertainty translates into modeling of processes such as flow, transport, wave, heat equations or even decisions made based on such models. There may be uncertainty related to the physics of these processes or other parameters that need to be specified to specify these processes. For example, solving partial differential equations requires boundary and initial conditions that may be uncertain.

3 Uncertainty assessment is subjective: while a "true" Earth exists with all of its true, but unknown properties, there is no "true uncertainty." The existence of a true uncertainty would call for knowing the truth, which would erase the need for uncertainty assessment. Uncertainty can never be objectively measured. Any assessment of uncertainty will need to be based on a model. Any model, whether statistically or physically defined, based on probability theory or fuzzy logic, requires implicit or explicit model assumptions (because of lack of knowledge or data), hence is necessarily subjective. There is no true uncertainty; there are only models of uncertainty, hence the title of this book.

4 High dimensional/spatial aspect: we are dealing with complex Earth systems that require a large amount of variables to describe them. Typically, we will work with gridded models to represent all aspects of the natural system. If each grid cell in a model contains a few variables, then easily we have millions of variables for even relatively small models. Standard approaches of probability become difficult to apply, since probability theory and statistical techniques common to most introductory text books has not been developed with these complex situations in mind. Often, it is necessary to perform some sensitivity analysis to determine which factors impact our decision most. Traditional statistical methods for sensitivity analysis are difficult to apply in this high dimensional and spatial context.

5 Several data sources informing various scales of variability: we will need to deal with a variety of data or information to constrain models of uncertainty. Without any data, there would be no modeling. Such data can be detailed information obtained from wells or more indirect information obtained from geophysical or remote sensing surveys. Each data source (such as wells) informs what we are modeling at a certain "volume support" (such as the size of a soil sample) and measures what we are targeting directly or indirectly, for example, electromagnetic (EM) waves for measuring water saturation.

Following this introductory chapter, this book covers many of these issues in the following chapters:

Chapter 2 Probability, Statistics and Exploratory Data Analysis: basically an overview of basic statistics and probability theory that is required to understand the material in subsequent chapters. The aim is not to provide a thorough review of these fields, but to provide a summary of what is relevant to the kind of modeling in this book.

Chapter 3 Modeling Uncertainty: Concepts and Philosophies: uncertainty is a misunderstood concept in many areas of science, so the various pitfalls in assessing uncertainty are discussed; also, a more conceptual discussion on how to think about uncertainty is provided. Uncertainty is not a mere mathematical concept, it deals with our state of knowledge, or lack thereof, as the world can be perceived by human beings. Therefore, it also has some interesting links with philosophy.

Chapter 4 Engineering the Earth, Making Decisions Under Uncertainty: the basic ideas of decision analysis are covered without going too much into detail. The language of decision analysis is introduced, structuring decision problems is discussed and some basic tools such as decision trees are introduced. The concept of sensitivity analysis is introduced; this will play an important role through many chapters in the book.

Chapter 5 Modeling Spatial Continuity: the chapter covers the various techniques for modeling spatial variability, whether dealing with modeling a rock type in the subsurface, the porosity of these rocks, soil types, clay content, thickness variations and so on. The models most used in practice for capturing spatial continuity are covered; these models are (i) the variogram/covariance model, (ii) the Boolean or object model and (iii) the 3D training image model.

Chapter 6 Modeling Spatial Uncertainty: once a model of spatial continuity is established, we can "simulate the Earth" in 2D, 3D or 4D (including time, for example) based on that continuity model. The goal of such a simulation exercise, termed stochastic simulation, is to create multiple Earth representations, termed Earth models, that reflect the spatial continuity modeled. This set of Earth models is the most common representation of a "model of uncertainty" used in this book. In accordance with Chapter 5, three families of techniques are discussed: a variogram based, object based and 3D training image based.

Chapter 7 Constraining Spatial Uncertainty with Data: this chapter is an extension of the previous chapter and discusses ways for constraining the various Earth representations or models with data. Two types of data are discussed: hard data and soft data. Hard data are (almost) direct measurements of what we are modeling, while soft data are everything else. Typically, hard data are samples taken from the Earth, while typical soft data are geophysical measurements. Two ways of including soft data are discussed: through a probabilistic approach or through an inverse modeling approach.

Chapter 8 Modeling Structural Uncertainty: the Earth also consists of discrete planar structures such as a topography, faults and layers. To model these we often use a modeling approach tailored specifically for such structures that is not easily captured with a variogram, object or 3D training image approach. In this chapter the basic modeling approach to defining individual faults and layers and methods of combining them into a consistent structural model are discussed. Since structures are often interpreted from geophysical data, the various sources of uncertainty for structural models and how they can be consistently constructed are discussed.

Chapter 9 Visualizing Uncertainty: because of the large uncertainty in Earth modeling and the many sources of uncertainty present, as well as the large amount of possible alternative Earth models that can be created, there is a need to get better insight into the integrated model of uncertainty through graphical representation. Some recently developed techniques based on distances to represent complex models uncertainty in simple 2D plots are discussed.

Chapter 10 Modeling Response Uncertainty: modeling uncertainty of the Earth by itself has little relevance in terms of the practical decision it is necessary to make. Instead, these models are used to evaluate certain scenarios or, in general, certain response functions, such as the total amount of contaminant, the best location for the sampling, the total amount of carbon dioxide that can be injected without risk, the best place to store nuclear waste and so on. This calls often for evaluating Earth models through CPU-expensive transfer functions, such as flow simulators, optimization codes, climate models and so on, that can take hours or days to run. A few techniques are presented for assessing response uncertainty that can deal, through model selection, with the issue of CPU cost in mind.

Chapter 11 Value of Information: before taking any data, such as costly drilling, sampling surveys or geophysical and remote sensing data, it can be quite useful to assess the value of taking such data. Such value will necessarily depend on the given model of uncertainty. Often, the more uncertain one is prior to taking any data, the more valuable the data may be. In this chapter, techniques are discussed for assessing such value of information in a formal decision analysis framework with a spatial context in mind.

Chapter 12 Case Study: the book concludes with a case study in value of information regarding a groundwater contamination problem. The aim of this case study is to illustrate how the various elements in this book come together: decision analysis, 3D modeling, physical modeling and sensitivity analysis.

Further Reading

BurVal Working Group (2006) Groundwater Resources in Buried Valleys: A Challenge for Geosciences. Leibniz Institute for Applied Geosciences (GGA-Institut), Hannover, Germany.

Jørgensen, F. and Sandersen, P.B.E. (2006) Buried and open tunnel valleys in Denmark – erosion beneath multiple ice sheets. *Quaternary Science Reviews*, **25** (2006) 1339–1363.

Huuse, M., Lykke-Andersen, H. and Piotrowski, J.A. (eds) (2003) Special Issue on Geophysical Investigations of Buried Quaternary Valleys in the Formerly Glaciated NW European Lowland: Significance for Groundwater Exploration. *Journal of Applied Geophysics*, **53**(4), 153–300.

2

Review on Statistical Analysis and Probability Theory

2.1 Introduction

This chapter should be considered mostly as a review of what has been covered in a basic course in statistics and/or probability theory for engineers or the sciences. It provides a review with an emphasis of what is important for modeling uncertainty in the Earth Sciences.

Statistics is a term often used to describe a "collection or summary of a set of numbers." Particular examples are labor or baseball statistics, which represent a series of numbers collected through investigation, sampling, or surveys. These numbers are often recombined or rearranged into a new set of clearly interpretable figures. Such summaries are needed to reduce the possibly large amount of data, make sense of the data, and derive conclusions or decisions thereupon. The scientific field of statistics is not very different. What statistics provides, however, is a rigorous mathematical framework in which to analyze sample data and to make conclusions on that basis. However, statistics as a traditional science has considerable limitations when applied to the Earth sciences. Key to modeling in the Earth Sciences is the spatial aspect or nature of the data. A sample or measurement is often attached to a spatial coordinate (x,y,z) describing where the sample was taken. Traditional statistics very often neglects this spatial context and simply works with the data as they are. However, from experience, it is known that samples located close together are more "related" to each other, and this relationship may be useful to us when interpreting our data. Geostatistics and also spatial statistics are fields that deal explicitly with data distributed in space or time and aim at explicitly modeling the spatial relationship between data.

Critical to any statistical study is to obtain a good overview of the data and to obtain an idea about its key characteristics. This analysis of the data is also termed exploratory data analysis (EDA). We will distinguish between graphical and numerical techniques.

Modeling Uncertainty in the Earth Sciences, First Edition. Jef Caers.
© 2011 John Wiley & Sons, Ltd. Published 2011 by John Wiley & Sons, Ltd.

2.2 Displaying Data with Graphs

In general, we consider two types of variables:

1 Categorical/Discrete Variables: gender, race, diamond quality category.

2 Continuous Variables: size, value, income, grade.

Some variables are naturally categorical, such as race, while other variables can be both categorical and continuous depending upon one's decision. For example, the quality of diamonds can be fair, good, excellent, or you can model it with the value of the diamond, which is a continuous variable expressed in US dollars (US$).

Both categorical and continuous variables exhibit a distribution. A distribution tells us how frequently a variable takes a given value. For categorical cases, this is straightforward. For each category, one records how many times that object falls in a particular category. For continuous variables, this is less obvious, and we will see methods to quantify frequencies for such variables. Note that continuous variables can always be categorized, but categorical variables cannot be made uniquely continuous.

2.2.1 Histograms

One of the most common methods for presenting data is the histogram. The histogram is a visual frequency table that presents how observed values fall within certain intervals (Figure 2.1). In a histogram, the continuous sample values are divided into intervals of usually equal width (bins). The width of these bins (i.e. the number of bins into which the range of values is divided) will determine how the histogram looks (Figure 2.2):

* Small bins – A lot of detail is observed but those details might be artifacts due to small sample fluctuations. One may lose the global overview of the message that the histogram brings, as it becomes too "noisy."

* Large bins – One might lose details and important features that are contained in the data.

Histograms are very useful for displaying data, because they are easy to understand and can be easily interpreted by those who are not specialized in statistics. However, later we will discuss more specialized techniques that allow better visualization of the data.

In terms of order of importance, one should look at the following when interpreting a histogram:

1 **The overall shape**: Shape can be symmetric or skewed. If skewed, it can be skewed to the left or to the right. Symmetric – left and right sides are almost, but not necessarily,

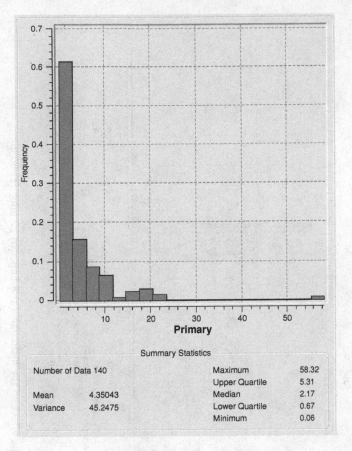

Figure 2.1 Example histogram.

perfect mirror images. Bell-shaped histograms are an example. Skewed – one side of the histogram extends further than the other side.

2 **Center, spread, skewness, mode**: The histogram allows one to visually approximate where the center of the data is. Mode – the mode is where the highest peak is. It represents the class of data that is most likely to occur. Two distinct modes could indicate two distinct populations in the data. Skewness – a measure of how far the histogram extends to the left or right. Spread – a measure of how much deviation there is from some average value.

3 **Outliers**: An outlier is an individual observation that seems not to belong to the pattern characterizing the overall distribution of the variable studied. A histogram is not a good tool to identify outliers, since it cannot determine the difference between an extreme value and an outlier. Extreme value – a legitimate sample from the population. It is not an anomaly but rather a high (or low) value that can regularly occur in a sample.

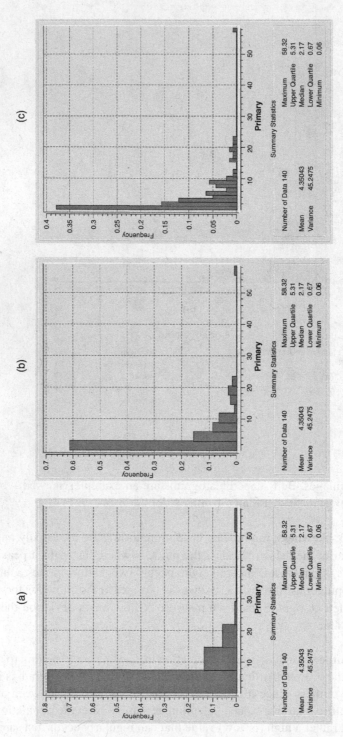

Figure 2.2   Example of bin sizes and the effect on the histogram (a) 8 bins, (b) 20 bins, (c) 50 bins.

2.3 Describing Data with Numbers

Histograms only provide a qualitative understanding of the data. We need to back up these visual observations with some informative summary values.

2.3.1 Measuring the Center

The mean is defined as the arithmetic average. If n observations $(x_1, x_2, x_3, \ldots, x_n)$, have been recorded then, the mean ($\overline{x}$) is defined as:

$$\overline{x} = \frac{x_1 + x_2 + \cdots + x_n}{n} = \sum_{i=1}^{n} \frac{x_i}{n}$$

Consider the following result from an actual diamond data set of 127 stones in which the value of each stone in US\$ is determined:

$$\overline{x} = \frac{x_1 + x_2 + \cdots + x_{127}}{127} = \text{US\$ } 142.9$$

Assume now that the values have been ranked from smallest value to largest value, so x_1 is the smallest and x_{127} the largest value. Consider now dropping the two smallest and two largest observations:

$$\overline{x} = \frac{x_3 + x_2 + \cdots + x_{125}}{123} = \text{US\$ } 65.3$$

Evidently, the mean is quite sensitive to extreme observations. In other words, the mean may not be a good summary of center when the distribution is skewed (more highs than lows or more lows than highs). To alleviate this, we use the median:

$$\text{Median}: \quad M = \text{middle value}$$

The median can be found by sorting all the data from the smallest to the largest observation, and then take the middle observation. Consider the following data set with seven observations ($n = 7$): 25, 5, 20, 22, 9, 15, 30.

Sorting the data set: 5, 9, 15, 20, 22, 25, 30, then the median = 20.

Now, consider a data set with eight observations (n = 8), in which seven are the same as above and the eighth is 16.

Sorting the data set: 5, 9, 15, 16, 20, 22, 25, 30.

What is done now? One can choose either 16 or 20 or take the average, which is 18. Note that the median can be 18 even though this value does not appear in the sample data.

The median is a more "robust" measure of the center than the mean with regards to outliers or extreme values. By more robust, we mean that the median is less sensitive to extreme values. We observed for the diamond example that removing the two smallest and two largest values can have a dramatic effect on the mean, but it has no effect on the median. In fact, the median does not depend much on the individual values recorded. It only depends on their relative ranking or ordering. If the distribution is symmetric, then the mean and the median will be close, and vice versa. If the mean and median are close to each other, then the data are most likely symmetric. For the above data set, the median is 18 and mean is 17.75. Thus, the data can be deemed as symmetric.

2.3.2 Measuring the Spread

The center is an important measure, but it does not tell the whole story of the data. For example, the average or mean pollution level on a regional scale, which is measured by monitoring at n stations, can be equal in two areas where one is conducting surveys. However, one cannot implement a remediation policy solely on the basis of the central values, since more extremely polluted sites might require more attention. This requires quantifying the variability or variation within a data set or population. A simple measure of spread could be to report the difference between the smallest and the largest values in a sample. But, evidently, this is not a robust measure of the spread.

Quartiles: Quartiles are quantiles (see later) just as the median in the sense that they are calculated in a similar way: the lower quartile is defined as a numerical value such that, out of n observations, 25% of these observations are lower or equal than the lower quartile and 75% are higher. The upper quartile is similar to the lower quartile except that now 25% of the observations are higher than the upper quartile. One can also state that the lower quartile is the median of the 50% smallest observations and that the upper quartile is the median of the 50% largest observations.

A good measure of spread is the interquartile range (IQR)

$$IQR = \text{upper quartile} - \text{lower quartile}$$

2.3.3 Standard Deviation and Variance

The most traditional way of summarizing data is by providing two numbers: the mean and the standard deviation. The standard deviation (s) and variance (s^2) look at the average spread in data, a spread that is measured around the mean.

$$x_1, x_2, x_3, \ldots, x_n \text{ is a set of } n \text{ observations.}$$

Spread around the mean:

$$x_1 - \overline{x} \longrightarrow (x_1 - \overline{x})^2$$
$$x_2 - \overline{x} \longrightarrow (x_2 - \overline{x})^2$$
$$\vdots$$
$$x_n - \overline{x} \longrightarrow (x_n - \overline{x})^2$$

Calculate the average:

$$s^2 = \sum_{i=1}^{n} \frac{(x_i - \overline{x})^2}{n - 1} = \text{(empirical) variance}$$

$$s = \sqrt{\sum_{i=1}^{n} \frac{(x_i - \overline{x})^2}{n - 1}} = \text{(empirical) standard deviation [has same units as } x\text{]}$$

Why divide by $(n - 1)$? We are not adding up n independent deviations. Indeed, we know that the sum of all deviations is equal to zero. But for all practical purposes, one can divide by n, because s^2 is a very variable number (it may varies substantially of another set of n data would be gathered) and n is not much different from $(n - 1)$ when n is large enough (larger than 10).

2.3.4 Properties of the Standard Deviation

The standard deviation (s) measures the spread around the mean $(\overline{x})$. Just as is the case of the mean, the standard deviation can be very sensitive to outliers. In fact, it is even more sensitive to outliers than the mean. The standard deviation is a good statistic to use for symmetric distributions. Otherwise, the IQR is a more robust measure of spread.

2.3.5 Quantiles and the QQ Plot

The term quantile is an important term in statistics. Median, lower quartile, and upper quartile are all examples of quantiles. A p-quantile with $p \in [0,1]$ (a percentile) is defined as that value such that a proportion of $100 \times p$ of the data does not exceed this value (in other terms: is lesser than or equal to this value).

For example: deciles = {0.1-quantile, 0.2-quantile, ..., 1-quantile}

Quantiles are useful in constructing so called quantile–quantile plots, which are graphical tools for comparing the distribution of two data sets. In a quantile–quantile plot, we compare the quantiles of two data sets. For example:

Data set 1	34	21	8	7	10	15
Data set 2	16	22	5	9	11	37

How many different quantiles can we calculate? In this case, we can calculate as many as we have data values. Indeed, we first arrange the data in numerical order.

Data set 1	7	8	10	15	21	34
Data set 2	5	9	11	16	22	37
Percentile	1/6	2/6	3/6	4/6	5/6	6/6

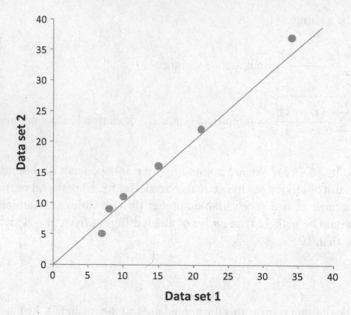

Figure 2.3   Example QQ plot.

A QQ plot is a cross-plot of the corresponding quantiles of the two data sets (Figure 2.3). QQ plots allow a better comparison between the distributions of two data sets than a histogram, particularly for skewed distribution and/or when only few data are available. One should observe a straight line of points on the 45° line to conclude that the data sets are comparable. Any deviation from that line may indicate differences between the two data sets. For example, if the two data sets have different mean values, then the QQ plot will still show a straight line, but this line will be parallel to the 45° line.

2.4 Probability

2.4.1 Introduction

Statistics uses the language of probability theory. This is essentially an entire field on its own, so we will only cover the basics. A probability is a purely mathematical concept and construction (one cannot observe a probability in nature). To illustrate this concept, consider the following statement:

There is a 60% probability/chance of finding iron ore in this region

What does this mean?
Interpretation 1: The geologist feels that, over the long run, in 60% of similar regions that the geologist has studied, iron ore will actually be found.

Interpretation 2: The geologist assesses, based and his/her expertise and prior knowledge, that it is more likely that the region will contain iron ore. In fact, 60/100 is a quantitative measure of the geologist's assessment about the hypothesis that the region will contain iron ore, where 0/100 means there is certainly no iron ore and 100/100 means there is certainly iron ore.

Interpretation 1 is also termed the frequency interpretation. Probability is interpreted as the ratio of success in the outcome of a repeated experiment, although in practice, such repetition need not be made explicit.

$$\text{Probability} = (\text{\# of successful events})/(\text{total number of trials})$$

For interpretation 2, probability is not thought of as being the property of an outcome (i.e., whether or not iron ore is actually found). Rather, it is considered to be a non-mathematical assessment based on prior knowledge by the person making the statement.

In this book, both interpretations are used. For example, we will present modeling techniques to figure out what the probability of contamination of a drinking well is given the available data. To do so, a number of alternative Earth models will be created, some of which show contamination some of them not; the frequency of contamination is interpreted as a probability. In the same context, however, geologist will need to decide what geological depositional system is present in order to create such Earth models. Consider that there is some discussion between a fluvial and an alluvial system being present. Since only one actual system is present, any probabilistic statement such as "the probability of the system being fluvial is x %" is not based on any frequency but on a personal assessment from the expert geologist based on his/her prior knowledge when interpreting the available data.

2.4.2 Sample Space, Event, Outcomes

Consider again the diamond example data set above. Taking a single stone from the deposit is considered the experiment; one also terms this "sampling" a stone. The outcome of the experiment is not known with certainty ahead of time. However, assume that the weight of a diamond is less than a big number, "BIG," but larger than zero. The set of all possible outcomes is also termed the sample space (S). For example:

$$\text{Sizes of Diamonds: } S = (0, \text{ BIG})$$
$$\text{Race: } S = \{\text{asian, black, caucasian, ...}\}$$

Any subset of this sample is termed an event:

$$\text{Event } E_1 = \{\text{a diamond of size larger than 5 ct}\}$$
$$\text{Event } E_2 = \{\text{a diamond of size between 2 and 4 ct}\}$$

The probability of an event occurring is denoted as $P(E)$. The so-called axioms of probability are then as follows:

Axiom 1: $0 \leq P(E) \leq 1$

Axiom 2: $P(S) = 1$

Axiom 3: $P(E_1 \cup E_2 \cup E_3 \ldots) = P(E_1) + P(E_2) + P(E_3) + \ldots$ if $E_1, E_2, E_3, \ldots$ are mutually exclusive events (which means E_i and E_j cannot both occur at the same time).

2.4.3 Conditional Probability

Conditional probability is probably the most important concept in probability theory for this book. In the notion of conditional probability, we ask everyday questions such as:

What is the probability of finding oil at location y given that oil has been found at location x?

What is the probability that site x is contaminated with lead given that one has sampled a concentration of y ppm lead at location z?

In general, we would like to quantify the probability that a certain event takes place, given that a piece of information is available or given that another event has taken place. For example:

Given that the Loma Prieta earthquake happened in 1989, what is the probability that a large earthquake will occur on the same fault in the next three decades?

The notation for conditional probabilities is:

$$P(\text{event } E \text{ occurs} \mid \text{event } F \text{ occurs}) = P(E \mid F)$$

In assessing conditional probabilities it is necessary to know whether events are related or not, otherwise, a conditional probability $P(E|F)$ is simply equal to $P(E)$.

Given that a regular coin in tossed nine times and observed heads all nine times, what is the probability of having heads the tenth time? The answer is $1/2$ since tossing events are assumed not related.

How do we calculate this conditional probability? To do so, we draw a Venn diagram (Figure 2.4).

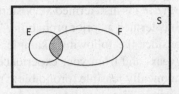

Figure 2.4   A Venn diagram displaying sample space and two events that are related.

Given that we know that E occurs, we can only be in circle E.

$$P(E) = \frac{\text{surface circle } E}{\text{entire box } S}$$

Given that F occurs, we can only be in circle F.

$$P(E\,|\,F) = \frac{\text{surface of intersection of circles } E \text{ and } F}{\text{surface of circle } F} = \frac{P(E \text{ and } F)}{P(F)}$$

2.4.4 Bayes' Rule

Bayes' rule or equation is one of the key concepts in probability theory and of paramount importance for the Earth Sciences. For example, the probability of finding a diamond in a region is an unconditional probability. As more data/information becomes available, (e.g., rock is kimberlitic, garnets have been found), the probability of an event will "change" either increasing or decreasing, that is, we are "learning" from the data. The magnitude of this change is governed by Bayes' rule.

Bayes' rule is simply deduced as follows:

1 $P(E \text{ and } F) = P(E\,|\,F)\,P(F)$

2 $P(E \text{ and } F) = P(F\,|\,E)\,P(E)$

Divide 1/2:

$$1 = \frac{P(E\,|\,F)\,P(F)}{P(F\,|\,E)\,P(E)}$$

Therefore:

$$P(E\,|\,F) = \frac{P(F\,|\,E)\,P(E)}{P(F)}$$

This equation provides a relationship between conditional and unconditional probabilities. If one conditional probability is known, then the other conditional probability can be calculated using Bayes' rule. $P(E|F)$ is also termed a "posterior" probability (after learning from the data), while P(E) is termed a "prior probability" (before having any data).

Why is this practical? Consider the following example: you have been working in the diamond industry for 30 years and from your experience you know that 1/10 of all deposits you studied are economically feasible (profitable). You are now given the study, as a consultant, to evaluate a newly discovered deposit. Your task is simply to report to the mining company the probability that the deposit is profitable. To help you with your work you analyze the garnet content of the deposit. Garnet is a mineral that tends to co-occur with diamonds. From your 30 years of experience in the industry and the database you kept on diamond deposits, you calculated that the probability of garnet exceeding 5 ppm for profitable deposits is 4/5 while this probability in the case of non-profitable deposits is only 2/5, Your analysis of garnet data for the current deposit reveals that the garnet content equal 6.5 ppm. What is the probability of profit for the current deposit? The solution to this problem is given by Bayes' rule.

E_1 = the deposit is profitable
E_2 = the deposit is not profitable
F_1 = the garnet concentration is more than 5 ppm
F_2 = the garnet concentration is less than or equal to 5 ppm

We need to determine $P(E_1|F_1)$. We know that the prior probability $P(E_1) = 1/10$ and, therefore, $P(E_2) = 9/10$, also:

$$P(F_1|E_1) = 4/5$$
$$P(F_1|E_2) = 2/5$$

To apply Bayes' rule we simply need $P(F_1)$, which can be found as follows:

$$P(F_1) = P(F_1|E_1)\,P(E_1) + P(F_1|E_2)\,P(E_2) = 4/5 \times 1/10 + 2/5 \times 9/10 = 22/50$$

The last statement is correct because $P(E_1) + P(E_2) = 1$, therefore:

$$P(E_1|F_1) = \frac{P(F_1|E_1)\,P(E_1)}{P(F_1)} = \frac{4/5 \times 1/10}{22/50} = \frac{2}{11}$$

This means that knowing the garnet concentration is above 5 ppm (the data) makes one about two times more certain that the deposit is profitable, indeed the ratio of posterior over prior is:

$$\frac{P(E_1|F_1)}{P(E_1)} = \frac{2/11}{1/10} = \frac{20}{11} = 1.81$$

2.5 Random Variables

At this point, we have studied two separate issues: (1) how to make numerical summaries of data, and (2) the study of "probability" in general without considering any data. In this section we will establish a link between the two – to try to quantify probabilities or other interesting properties from the data set. A key link will be the concept of the random variable. A random variable is a variable whose value is a numerical outcome of a random experiment. A random variable is not a numerical value itself. It can take various outcomes/values, but we do not know, in advance, exactly which value it will take. Examples are rolling a dice, drawing a card from a deck, sampling a diamond stone from a diamond deposit. All of these are variables that can be described by a random variable.

We will use as a capital letter such as X or Y to denote a random variable. The capital letter is important, because we use it to indicate that the value is unknown. The outcome of a random variable is then denoted by a small letter such as x or y.

$P(X \leq x)$: denotes the probability that the random variable X is smaller than a given outcome x. Recall that "$X \leq x$" is termed an event.

2.5.1 Discrete Random Variables

A random variable that can take only a limited set of outcomes or values is termed a discrete random variable. An example is rolling a dice: there are only six possible outcomes. The frequency at which the outcomes occur or the way the random variable is distributed can be described by a probability mass function using the following notation:

$$p_X(a) = P(X = a)$$

for dice: $P(X = 1) = p_X(1) = 1/6$; $P(X = 2) = p_X(2) = 1/6$, and so on. Note the notation $p_X(a)$, which means that we evaluate the probability for random variable X to take the value a.

2.5.2 Continuous Random Variables

While in the discrete case, the frequency of possible outcomes can be counted, the number of possible outcomes cannot be counted for a continuous random variable. In fact, there are an infinite number of possibilities. Because of this, P(diamond size = 1 ct) = 0 because there are an infinite (at least theoretically) possibilities, hence any number divided by infinite is zero. There are two ways to describe the possible variations of a continuous random variable. Both ways are equivalent. (1) probability density function (pdf) and (2) cumulative distribution function (cdf).

2.5.2.1 Probability Density Function (pdf)

The probability density function, which we denote as $f_X(x)$, is defined as an integral of a positive function and this integral (surface area) denotes a probability:

$$P(a \leq X \leq b) = \int_a^b f_X(x)dx$$

This seems a very contorted way to define a probability, but we need to do it in this mathematical way for continuous variables because of the 1/infinite reason mentioned above. The notation $f_X(x)$ now also becomes a bit more clear. The function f describing the probabilistic variation of random variable X is evaluated in the point x.

Some important properties are

$$\int_{-\infty}^{+\infty} f_X(x)dx = 1 \qquad \text{"some outcome will occur for sure"}$$

$$f_X(x) \geq 0 \qquad \text{"probabilities cannot be negative"}$$

$$P(X = x) = 0 \qquad \text{as discussed above}$$

Does the function value $f_X(x)$ have any meaning? It certainly does not have the meaning of a probability. It really only has meaning when comparing two outcomes, x_1 and x_2. Then the ratio:

$$\frac{f_X(x_1)}{f_X(x_2)}$$

denotes how many times more (or less) likely the outcome x_1 will occur compared to x_2. Note that the term "likely" (also used as likelihood) is not the same as "probability". For example if the ratio is four, as the case in Figure 2.5, then x_1 is four times more likely to occur than x_2. Note that this is not the same as saying "four times more probable."

2.5.2.2 Cumulative Distribution Function

A completely equivalent way of describing a random variable is a cumulative distribution (Figure 2.6):

$$F_X(x) = P(X \leq x)$$

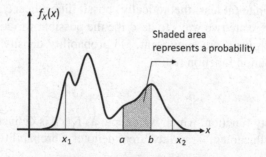

Figure 2.5   Example of a probability density function.

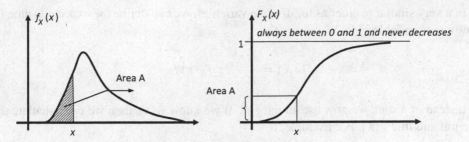

Figure 2.6 Definition of a cumulative distribution function.

The relationship between $F_X(x)$ and $f_X(x)$ is:

$$F_X(x) = \int_{-\infty}^{x} f_X(y)dy \Rightarrow f_X(x) = \frac{dF_X(x)}{dx}$$

2.5.3 Expectation and Variance

2.5.3.1 Expectation

Firstly, consider a discrete random variable X with probability mass function $p_X(x)$. X has K possible outcomes, say $x_1, x_2, x_3, \ldots, x_k$:

$$x_1 : P(X = x_1) = p_X(x_1)$$
$$x_2 : P(X = x_2) = p_X(x_2)$$
$$x_3 : P(X = x_3) = p_X(x_3), \text{ etc.}$$

The expected value, using the notation E[X], is defined as:

$$E[X] = \sum_{k=1}^{K} x_k P(X = x_k)$$

For example: in rolling a dice we have:

Possible outcomes : $x_1 = 1, \ x_2 = 2, \ x_3 = 3, \ x_4 = 4, \ x_5 = 5, \ x_6 = 6$

$$E[X] = 1 \times \tfrac{1}{6} + 2 \times \tfrac{1}{6} + 3 \times \tfrac{1}{6} + 4 \times \tfrac{1}{6} + 5 \times \tfrac{1}{6} + 6 \times \tfrac{1}{6} = \tfrac{7}{2}$$

Apparently, the expected value of X need not be a value that X could assume. So E[X] is not the value that one "expects" X to have, but rather E[X] is the average value of X in a large number of repetitions of the experiment.

In a very similar manner as for discrete variables, we can define the expected value for continuous variables:

$$E[X] = \int_{-\infty}^{+\infty} x\, f_X(x)dx$$

Instead of a sum, we now use an integral. If we know $f_X(x)$, then we can calculate this integral and find E[X]. For example, if

$$f_X(x) = \frac{1}{\sqrt{2\pi}\sigma} \exp\left(-\frac{1}{2}\left(\frac{x-\mu}{\sigma}\right)^2\right)$$

then with some calculus this will result in:

$$E[X] = \int_{-\infty}^{+\infty} x\, \frac{1}{\sqrt{2\pi}\sigma} \exp\left(-\frac{1}{2}\cdot\left(\frac{x-\mu}{\sigma}\right)^2\right) = \mu.$$

2.5.3.2 Population Variance

In a sense, the expected value of a random variable is one way to summarize the distribution function of that variable. So how do we summarize the spread of that population? Equivalently, as we did for the data (where we used the empirical standard deviation), we also use a measure – the population variance:

$$\mathrm{Var}[X] = E\left[(X - E[X])^2\right] = \int_{-\infty}^{+\infty} (x-\mu)^2 f_X(x)\, dx$$

If we use the same function above, then:

$$\mathrm{Var}[X] = \int_{-\infty}^{+\infty} (x-\mu)^2 \frac{1}{\sqrt{2\pi}\sigma} \exp\left(-\frac{1}{2}\left(\frac{x-\mu}{\sigma}\right)^2\right) dx = \sigma^2$$

2.5.4 Examples of Distribution Functions

2.5.4.1 The Gaussian (Normal) Random Variable and Distribution

The Gaussian or Normal distribution is a very specific distribution that has the following mathematical expression:

$$f_X(x) = \frac{1}{\sqrt{2\pi}\sigma} \exp\left(-\frac{1}{2}\left(\frac{x-\mu}{\sigma}\right)^2\right)$$

with μ = population mean or expected value; σ = population standard deviation. Figure 2.7 shows some examples with various values for μ and σ.

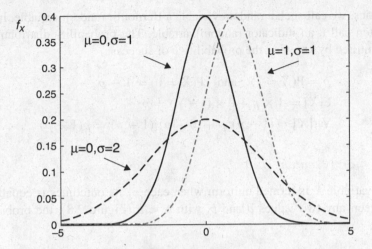

Figure 2.7 Gaussian distribution for various values of μ and σ.

The Gaussian distribution has two parameters μ and σ that can be freely chosen (remembering that $\sigma > 0$!). The parameter μ "regulates" the center of the distribution. It is the mean if you have an infinite amount of samples from a random variable X with this distribution. The parameter σ "regulates" the width of the bell-shaped curve.

Other properties of this distribution function are:

- population mean = population median

- population mode = population mean

- There is no mathematical expression for $F_X(x)$. You need to use a computer or a table from a book.

2.5.4.2 Bernoulli Random Variable

The simplest case of a discrete random variable is one that has two possible outcomes. For simplicity, we will call these categories 0/1.

$$
\begin{aligned}
X &= 0 &&\text{If a trial results in "failure"} \\
X &= 1 &&\text{If a trial results in "success"}
\end{aligned}
$$

A trial should be treated in the broadest sense possible.
Examples:

Finding a diamond larger than 2 ct means "success."

Rolling ones twice in a row means "success."

Finding a diamond smaller than 4 ct means "success."

In statistics, we call such a random variable a Bernoulli random variable. In geostatistics, we often call it an indicator random variable. The probability distribution is completely quantified by knowing the probability p of success:

$$p = P(X = 1) \quad \text{and} \quad P(X = 0) = 1 - p$$
$$E[X] = 1 \times p + 0 \times (1 - p) = p$$
$$\text{Var}[X] = (1 - p)^2 p + (0 - p)^2(1 - p) = p(1 - p)$$

2.5.4.3 Uniform Random Variable

A random variable X is termed uniform when each of its outcomes is equally likely to occur between any two values a and b, with $a < b$ (Figure 2.8), the probability density equals:

$$f_X(x) = \begin{cases} 1/(b - a) & \text{if } a \leq x \leq b \\ 0 & \text{elsewhere} \end{cases}$$

The uniform random variable is important in the context of generating random numbers on a computer.

2.5.4.4 A Poisson Random Variable

Examples for which a Poisson random variable is appropriate:

- The number of misprints on a page of a book

- The number of people in a community that are 100 years old

- The number of transistors that fail in their first day of use.

Poisson random variables typically have the following characteristics:

$p = $ probability of the event occurring is small
$n = $ number of trials is large

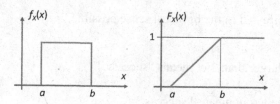

Figure 2.8   Uniform pdf and cdf.

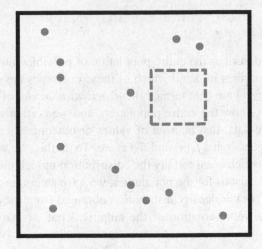

Figure 2.9   Random distribution of points over an area.

In the Earth sciences, the Poisson distribution is important because it has a spatial connection. Take an area with certain objects (diamonds, trees, plants, earthquakes). Take a small box and put it inside this area, as done in Figure 2.9. The random variable describing the number of points that you will find in the box is a Poisson random variable and follows the following equation:

$$p_X(i) = P(X = i) = e^{-\lambda}\frac{\lambda^i}{i!} \quad \lambda = \text{average number of points in the box}$$

In Chapter 5, we will discuss Boolean or object models, where we will simulate objects in space. To do so we will make use of the Poisson process, that is, the process of spreading objects randomly in space as done in Figure 2.9. Note that the coordinate X and Y of each point are also random variables, namely uniform random variables.

2.5.4.5 The Lognormal Distribution

A variable X is lognormally distributed if and only if $\log X$ is normally/Gaussian distributed. So, if we calculate the log of the data, then the histogram should look like a normal distribution in case that variable is lognormally distributed. The lognormal distribution has, therefore, two parameters – mean and variance. The lognormal distribution can be extremely skewed. Hence, it is an ideal candidate for describing skewed data sets. The lognormal variable is also positive. This makes it a very useful distribution for most Earth Science data which are strictly positive; permeability (Darcy), magnitudes, and grain sizes (mm), for example, are often lognormal.

2.5.5 The Empirical Distribution Function versus the Distribution Model

A random variable X describes the entire population of possible outcomes, and its distribution (pdf or cdf) describes in detail which of these outcomes are more likely to occur than others. $F_X(x)$ or $f_X(x)$ are also termed the *distribution model* of the population. Unfortunately, we do not know the entire population, and we certainly do not know $F_X(x)$ or $f_X(x)$. We only have data, that is, a set of values or outcomes of sampling. Using the data, we will have to guess what $f_X(x)$ and $F_X(x)$ are. To do this, we will use the empirical distribution function, which is essentially the "distribution model of the data" and not the entire population of X. Just as for the population, we have an empirical pdf and cdf.

Empirical pdf $= \hat{f}_X(x) =$ density distribution obtained from the data. The histogram is, in fact, a graphical representation of the empirical pdf, so we will call $\hat{f}_X(x)$ the histogram.

Empirical cdf $= \hat{F}_X(x) =$ the cumulative distribution function based upon the data. It is constructed as shown in Figure 2.10.

- Sort the data and plot them on the x-axis.

- A cumulative probability specifies the probability of being below a threshold, so for the empirical cdf this becomes:

$$\hat{F}_X(x) = P(X \leq \text{observed datum } x)$$

$P(X \leq x_1) = 1/6$ *then 16%* of the data is less than or equal to $x_1 = 3.2$.

$P(X \leq x_2) = 2/6$ *then 33%* of the data is less than or equal to $x_2 = 8.6$.

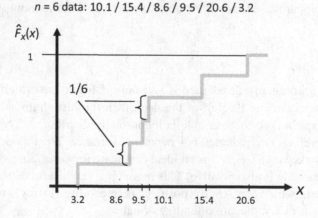

Figure 2.10   Empirical cdf.

2.5.6 Constructing a Distribution Function from Data

An important task in statistical analysis is to figure out which distribution is suitable to model the data: is it normal? Lognormal? Uniform? Unfortunately, no set of distribution models exists that is "flexible" enough to fit all the data sets that are observed in nature. Many of the theoretical distribution models (such the normal and lognormal) stem from an era where computers were not available and modelers used functions that had only a few parameters they could easily estimate. In this book, a more computer-oriented method of interpolation/extrapolation that uses the data themselves to construct a distribution model is advocated. Our anchor point will be the empirical distribution.

Figure 2.11 provides an example, where we assume the data is bounded between 0 and 100. Recall that in the empirical cumulative distribution, we list the data $x_1, \ldots, x_6$, and

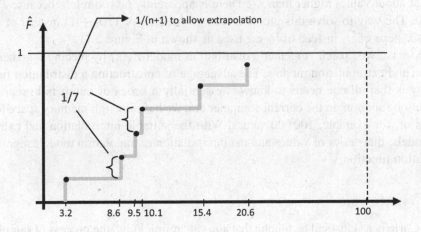

6 data samples: 3.2 / 8.6 / 9.5 / 10.1 / 15.4 / 20.6

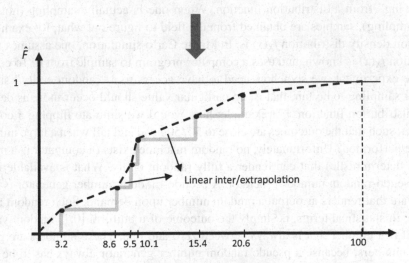

Figure 2.11 Empirical cdf for building a distribution model directly from data.

then we make steps of $1/n$ between them. However, when we want a distribution model for the entire population, we need to perform two additional steps:

- **Interpolate**: We need to know what happens between x_2 and x_3, or x_3 and x_4 and so on.

- **Extrapolate**: We need to know what happens for observations larger than x_6 and smaller than x_1. Indeed, the data are only a limited sample of the entire population. In the entire population it can be expected that some values are large than x_6 and some are smaller than x_1.

Therefore, we "complete" the empirical distribution by introducing so-called interpolation and extrapolation models, which can be chosen by the modeler, for example a linear or parabolic/hyperbolic type function. There is no problem to extrapolate lower than x_1, but what about values higher than x_6? There is apparently no room left, because $F_X(X > x_6) = 0$. The way to solve this problem is to go in steps of $1/(n+1)$ instead of $1/n$. In this case, steps of $^1/_7$ instead of $^1/_6$ are used as shown in Figure 2.11.

In essence, we "patch" together a distribution model $F_X(x)$ by piecing together interpolation and extrapolation models. The advantage of constructing a distribution function this way is that all one needs to know is essentially a series of numbers to represent a distribution function. In the current computer era, we have enough memory space to store a series of, for example, 100 000 values. With the suitable interpolation and extrapolation models, that series of values and the interpolation/extrapolation models represents a distribution function.

2.5.7 Monte Carlo Simulation

Monte Carlo is a statistical technique that aims at "mimicking" the process of sampling an actual phenomenon. Therefore, Monte Carlo simulation is often referred to as "sampling" or "drawing" from a distribution function. When one is actually sampling (not Monte Carlo sampling), samples are obtained from the field to figure out what, for example, the population density distribution $f_X(x)$ is. In Monte Carlo simulation, one assumes that the distribution $f_X(x)$ is known, and uses a computer program to sample from it. To construct a sample experiment, we somehow need to have access to a "random entity," since we want our sampling to be fair, that is, no particular value should occur more as described by the distribution function. For example: how would we simulate flipping a coin on a computer, such that the outcomes are close to 50/50 head and tail when a large number of trials are performed? Unfortunately, no random machine exists (a computer is a machine and still deterministic) that can render a fully random entity. What is available is a so-called pseudo-random number generator. A pseudo-random number generator is a piece of software that renders as output a random number upon demand. Such random number or value, in statistical terms, is simply the outcome of a uniform [0,1] random variable. Hence, it is a number that is always between zero and one. These numbers are pseudo-random numbers, because a pseudo random number generator always has to be started with what is called a "random seed." For a given random seed, one will always obtain the

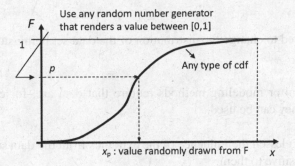

Figure 2.12   Monte Carlo simulation; a computer program generates p, from which x_p is obtained

same sequence series of random numbers. For example, using MATLAB on the computer currently on my desk, using a random seed of 69 071 and the random number generator implemented in MATLAB the following list of random numbers is obtained: 0.10135, 0.58382, 0.98182, 0.0534, 0.48321, 0.65310, and so on. Using the same machine and software, one will always get the same sequence for a given random seed.

We can now generate samples from a specified distribution $f_X(x)$ as follows (Figure 2.12):

1 Draw a random number.

2 Use the cumulative distribution to "look up" the corresponding sample value.

3 Repeat this as many times as samples are needed.

x_p is then a sample value of $F_X(x)$ obtained by Monte Carlo simulation. If this experiment is repeated a large number of times (trial, samples or drawings), then one would get an empirical distribution of the x_p values that approximates the $F_X(x)$ from which you sampled. This makes sense: if we sample a large number of values x_p, construct the empirical cdf with steps of $1/n$ and plot it on top of the graph in Figure 2.12, then this step function will approximate the actual cdf.

Why do we use Monte Carlo experiments and samplers?

- It is used to create data sets on which one can try out certain computer experiments or test methodologies.

- It can help predict the effects of doing certain sampling campaigns and helps in the design of sampling surveys.

- It is used to create Earth Models and model uncertainty, as will be extensively discussed in Chapters 5–8.

2.5.8 Data Transformations

Often, one will need to change the distribution of the data set being studies. This is necessary because:

- Certain statistical or modeling methods require that data are, for example, standard normal before they can be used.

- One may want to lessen the influence of the extremes from the data such that estimates become less sensitive to them.

How is this done? Consider a data set containing values:

$$x_1 = 8; \ x_2 = 3; \ x_3 = 6; \ x_4 = 9; \ x_5 = 20;$$

One would like transform these values into five standard normal values: y_1, y_2, y_3, y_4, y_5. This is done as follows:

1 Convert the five values (x_1 through x_5) into uniform values (r_1 through r_5): simply rank them from low to high and associate $1/6$ (or $(1/n + 1)$ in general) with each value. Call these ranked values x_r, then $x_{r,1} = 3; x_{r,2} = 6; x_{r,3} = 8; x_{r,4} = 9; x_{r,5} = 20;$

2 Using the same techniques as in Monte Carlo simulation, graphically find the corresponding standard normal distribution using the steps $1/(n + 1)$. Figure 2.13 shows how a so-called "normal score transformation" proceeds. In a normal score transformation, the y values have a standard normal distribution.

A back transformation then operates the opposite way. For example in Figure 2.13 one selects a standard Gaussian value y_s and finds the corresponding value x_s. This back transformation requires the interpolation and extrapolation functions that were introduced in Figure 2.11.

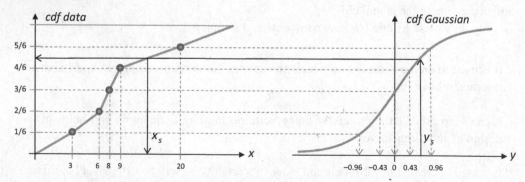

Figure 2.13   Transformation of values x into standard Gaussian values, y.

2.6 Bivariate Data Analysis

2.6.1 Introduction

Many statistical methods/problems involve more than one variable. In bivariate statistics, the association between two random variables is studied. For example, the occurrence of garnet minerals might indicate potential diamond presence. Hence, the random variable "ppm garnet" will show correlation with the random variable "ct diamond." They are not independent.

In bivariate statistics we explore and model these relationships: a typical question would be how one can know something about random variable X_1 if we have information on X_2. If these variables are correlated, then one variable necessarily contains information about the other. Bivariate analysis is not only used for studying the association between two variables (for example length and width of a fossil) but also two time events:

$X_1 =$ the temperature tomorrow

$X_2 =$ the temperature today.

Such studies are also termed time series analysis. Instead of looking at a random variable distributed in time, the variable can be distributed in space. Then, we are studying a variable at a certain location, and we can compare it with a variable at a given distance apart in space. This will be the topic in Chapter 5. For example, if one observes x ppm of gold at location y, does this inform about gold at location z, which is a distance of 10 ft away? We know that gold does not occur randomly, so the answer to this question is likely positive. Gold has some underlying geological structure (e.g., veins) that makes the observations dependent. Hence, we can exploit this dependency to predict the concentration of gold at location z from the observation at location y.

2.6.2 Graphical Methods: Scatter plots

As in univariate statistics, the variables can be continuous, discrete, or a combination of the two. When looking at two variables, we need a concurrent sample of both.

X_1 variable 1 : $x_{11}, x_{12}, x_{13}, \ldots, x_{1n}$

X_2 variable 2 : $x_{21}, x_{22}, x_{23}, \ldots, x_{2n}$

For example, consider an actual data set from a kimberlite deposit with the following: variable 1 = diamond size (ct), variable 2 = diamond value ($) (Figure 2.14). Clearly an association exists between the two variables, but the association is not straightforward. A large diamond does not always have higher value than a smaller diamond. The confounding factor here is quality, namely, diamonds with better quality are more desirable even if they are smaller.

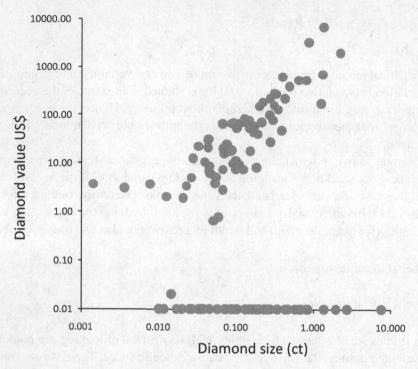

Figure 2.14 Diamond values in US$ vs diamond size in karat; note the log scale on both axes.

A scatter plot shows the relationship between two variables. In interpreting such a plot we look at the following in order of importance:

- Clustering: are there various groups of points?

- Strength of association: within such groups are the points randomly spread or more organized?

- Trend or sign of association: is the cloud of such group tilt upward or downward?

- Shape of association: what is the shape of a cluster?

Consider the example in Figure 2.15. How is this plot analyzed?

1 Look for clustering in the scatter plot: for example, look for lumps of data points that lie together. These clusters usually have a clear physical explanation to them (two different populations or deposits for example). However, be careful if a physical explanation cannot be found – often it looks like there are two clusters, but in reality this is simply due to the sampling fluctuations or sheer luck/misfortune. Two clusters are interpreted in Figure 2.15.

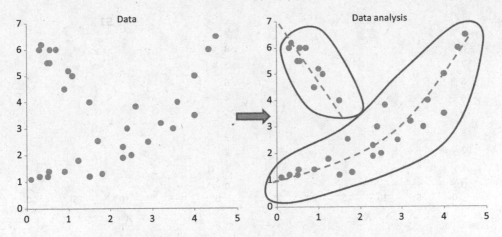

Figure 2.15 Example of a bivariate data set and its analysis.

2 For each cluster look at strength of association: the stronger the association, the smaller is the spread within a cluster. The smaller cluster in Figure 2.15 seems that have stronger association than the larger cluster.

3 Trend: variables are positively or negatively associated. When one variable increases and the other variable also increases = positive association. Clearly, the smaller cluster has negative association, while the larger cluster has positive association.

4 Shape: Look at the pattern of the points: do they lie more or less on a straight line or can one detect curve-linear variation in the association? When points are close to a straight line behavior, then the relationship is called linear.

Note: graphical summaries should still be made for each variable separately. This means looking at histograms, quantile plots, and so on of each variable separately.

2.6.3 Data Summary: Correlation (Coefficient)

2.6.3.1 Definition

Summarizing a scatter plot such as Figure 2.15 with easily interpretable numbers is not an easy task. The task becomes easier when the plot shows a linear association as occurs in Figure 2.16. This is the case, because there is only one cluster of points. Then, we have a measure of the strength of association of the two variables = correlation coefficient (r):

$$r = \frac{1}{n-1} \sum_{i=1}^{n} \left(\frac{x_i - \overline{x}}{s_x} \right) \cdot \left(\frac{y_i - \overline{y}}{s_y} \right)$$

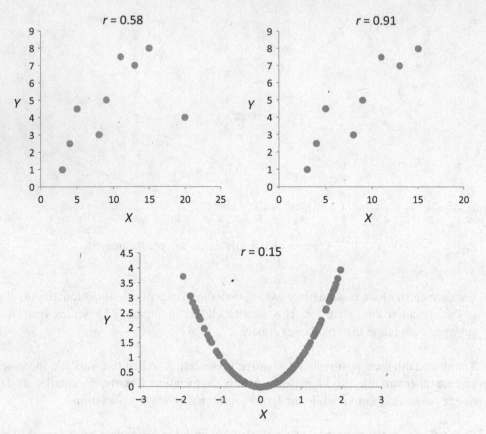

Figure 2.16 Example data sets with respective correlation coefficients showing that r is sensitive to outliers and is only meaningful for linear correlation.

Recall that:

$$s_x = \sqrt{\frac{1}{n-1} \cdot \sum_{i=1}^{n} (x_i - \overline{x})^2}$$

Analyzing this equation, define:

$$\tilde{x}_i = \frac{x_i - \overline{x}}{s_x} \quad \text{and} \quad \tilde{y}_i = \frac{y_i - \overline{y}}{s_y}$$

These are termed "standardized values" in the sense that the mean of these values is zero and their standard deviation unity:

$$r = \frac{1}{n-1} \sum_{i=1}^{n} \tilde{x}_i \cdot \tilde{y}_i$$

r is a measure of the strength and direction of the linear relationship. Indeed, if x and y tend to be both positive most of the time than r will have a larger value than when x and y have opposite signs most of the time. In fact:

If $|r|$ larger: stronger correlation

If $r > 0$: positive correlation

If $r < 0$: negative correlation

If $|r| = 1$: perfect linear correlation

If $r = 0$: no linear correlation

Also, the range of r is restricted to $[-1,1]$.

2.6.3.2 Properties of r

The following important properties of r should be noted.

1 r is scale independent. It does not matter whether the variables are measured in feet or centimeters.

2 Correlation measures only a linear relationship. If r is close to zero, then this does not necessarily mean that there is no relationship between X and Y (Figure 2.16).

3 r is strongly affected by a few outlying observations (Figure 2.16).

Further Reading

Borradaile, G.J. (2003) *Statistics of Earth Science Data*, Springer.

Davis, J.C. (2002) *Statistics and Data Analysis in Geology*, John Wiley & Sons, Inc.

Rohatgi, V.K. and Ehsanes Saleh, A.K. Md. (2000) *An Introduction to Probability and Statistics*, John Wiley & Sons, Inc.

3

Modeling Uncertainty: Concepts and Philosophies

Imagination is more important than knowledge: for knowledge is limited to what we know and understand while imagination embraces the entire world and all that ever will be known and understood

—Albert Einstein

3.1 What is Uncertainty?

Uncertainty is present in many aspects of scientific work, engineering as well as everyday life. Each field uses different terminologies and ways to describe, quantify and assess uncertainty. In this chapter, important concepts are discussed, even some philosophical approaches to "uncertainty" that can be treated purely scientific, but because of its many applications has also a societal impact. The discussion in this chapter is quite general and may apply to many fields; the chapter is concluded by applying these concepts to some real cases to illustrate more concretely what they mean.

What is uncertainty? An often approachable way to look uncertainty is to consider causality: "uncertainty is caused by an incomplete understanding about what we like to quantify." Quantifying uncertainty is not trivial. One may be tempted to state that quantifying what we don't know is the opposite of quantifying what we know. So, "once I know what I know, I also know what I don't know"; unfortunately not. Quantifying what one doesn't know is an inherently subjective task that cannot be tested against an irrefutable truth, as is discussed at length in this chapter. As such we will see that there is no true uncertainty, in other words we will never know whether a quantification of uncertainty is either correct or the best possible. For example, in aquifer modeling, while there is a "true" but unknown aquifer with all of its true, but unknown geological, geophysical and hydrological properties, there is, however, no "true uncertainty" about the lack of understanding of this aquifer. The existence of a true uncertainty would call for knowing the true aquifer, which would erase the need for uncertainty assessment. Un-

Modeling Uncertainty in the Earth Sciences, First Edition. Jef Caers.
© 2011 John Wiley & Sons, Ltd. Published 2011 by John Wiley & Sons, Ltd.

certainty can never be objectively measured, as a rock type or an elevation change could be perfectly measured if perfect measurement devices were available. Any assessment of uncertainty will need to be based on some sort of model. Any model, whether statistically or physically defined, requires implicit or explicit model assumptions, data choices, model calibrations and so on, which are necessarily subjective. A preliminary conclusion would therefore be that there is no true uncertainty; there are only models of uncertainty.

For example, the quantity "60%" in the statement: "given the meteorological data available, the probability of raining tomorrow is 60%" can never be verified objectively against a reference truth, because the unique event "it rains tomorrow" either happens or does not happen. There exists no objective measure of the quality or "goodness" of a choice or decision made prior to knowing the result or outcome of that decision. Nor does acquiring more observations necessarily guarantee a reduction of uncertainty. The acquisition of additional observations may result in a drastic change of our understanding of the system, it changes our interpretation of what we don't know (or in hindsight, didn't know). It just means that our initial model of uncertainty (prior to acquiring additional observations) wasn't realistic enough.

3.2 Sources of Uncertainty

Various taxonomies of sources of uncertainty exist. Those sources most relevant to the applications in this book are discussed. At the highest level one can distinguish uncertainty due to process randomness and uncertainty due to limited understanding of such processes. Consider this in more detail:

- **Process randomness**: due to the inherent randomness of nature, a process can behave in an unpredictable, chaotic way. Examples are the behavior of clouds, the turbulent flow in a pipe or the creation of a hurricane in the Atlantic Ocean. Literally the wing flap of a butterfly (termed the butterfly effect) off the coast of West Africa can create a major hurricane hitting the USA. Also, this type of uncertainty is particularly relevant to studying human behavior, societal and cultural tendencies and technological surprises. This uncertainty is less relevant to this book, where we study mostly physical processes that are considered deterministic (such as flow in a porous medium); however, many of the techniques described in this book could account for such uncertainty.

- **Limited understanding**: this source is related to the limited knowledge of the person performing the study or modeling task in question.

Limited understanding is therefore the main focus here. In this category, many subcategories can be classified. Some, as will become clear, are more relevant to the topic of this book.

- **"We roughly know"**: this uncertainty refers to the so-called "measurement error". Pretty much every physical quantity we measure is prone to some random error. Sometimes this error is negligible compared to other sources of uncertainty; for example, we can measure the porosity of a core sample reasonably accurately; other measurements may have considerable and relevant error.

- **"We could have known"**: rarely can we sample a phenomenon exhaustively. Since many phenomena may vary extremely in space and time, any lack of measurements over that space–time continuum will lead to uncertainty.

- **"We don't know what we know"**: Different data sets or observations can be interpreted differently by different field experts, each reaching a possible large variety of conclusions.

- **"We don't know what we don't know"**: this uncertainty is related to processes or phenomena we cannot even imagine being present or plausible. We do not observe them, nor do we think they are theoretically or practically possible. Sometimes this uncertainty is termed the epistemological uncertainty. Epistemology refers to theory of knowledge, an area of philosophy that has many applications in science and society.

- **"We cannot know"**: uncertainty related to the fact that some phenomena can never be measured since they are too far, for example, the properties of the inner core of the Earth.

3.3 Deterministic Modeling

In many practical cases, deterministic modeling still prevails. It is, therefore, important to describe why this is the case and what the role of such deterministic modeling is. Deterministic modeling refers to the building, construction of one single Earth model, whether this model is geological, physical, and chemical or any combination thereof, whether the model is 3D and/or varies over time. For example in the Danish aquifer case (Chapter 1), a geophysicist can take the result of a processed time-domain electromagnetic surveys (TEM) survey (such processing relies heavily on the physics of electromagnetic induction waves) (Figure 3.1) and create a deterministic contouring of the channel boundaries.

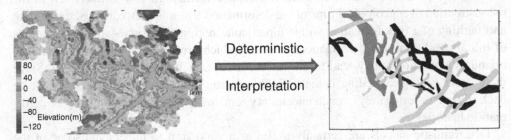

Figure 3.1 Processed TEM image (left) and deterministic geological interpretation of buried valleys (right).

In some cases, one may built a few 3D models (3–5), for example based on different scenarios for the geological setting, or simply by running a stochastic method such as in Chapters 5–8 a few times; in the Danish case, this could include modifying the buried valley positions slightly to account for the difficulty in interpreting accurately the TEM image in Figure 3.1. Such an approach is, however, not necessarily a rigorous exercise in modeling uncertainty, the idea here is just to explore a few alternative models. In the case of Figure 3.1, any rigorous uncertainty assessment would not only include uncertainty in the contouring but also uncertainty in processing the raw TEM observations to create the TEM image data set in Figure 3.1. The relationship between raw observations and data sets is discussed later in this chapter.

A single deterministic model is often built simply because of time (CPU or man-hour) constraints. Such a model may contain "all bells and whistles" that the model builder believes is representing those processes occurring in nature. An example of a deterministic model is a process-based model; examples are discussed in Chapter 5. This model explicitly simulates the dynamic process that took place or will take place. Examples of such process models are general circulation models (GCMs) for studying climate change, the simulation of sedimentary deposition to represent an aquifer or oil reservoir, the mechanical simulation of fracture growth and the simulation of carbonate reef growth. Deterministic models therefore aim, in the first place, to be physically realistic, that is, represent realistically physical and dynamic processes as well as their mutual interaction. In this sense, deterministic models are often deemed superior to models that are physically less realistic (simpler models). If that is the case then one should define specifically what is meant by "superior". Also, even a fully theoretical model is often dependent on *ad hoc* or arbitrary adjustments of physical parameters. This is certainly true for 3D gridded models, where the size of the grid cell may be large (km); therefore, any fine scale features or processes may need to be aggregated (i.e., summed/integrated) into a large grid cell. Indeed, a reservoir can never be represented by modeling every sand grain contained in it or a climate model by including every single cloud formation. Hence, even a full deterministic physical model may heavily rely on empirical or *ad hoc* determined input parameters, hence subject to a great deal of uncertainty in its physical representation, which is supposed to be its strength.

In addition, physical models often require initial conditions, boundary conditions that may be prone to large uncertainty. This is the case in modeling of the tectonic deformation of a particular region. In modeling such deformation, one is interested in the reconstruction over geological time of the deformation that took place, that is, the folding and faulting of a rock formation within a particular region or basin. Many combinations of initial conditions and deformation histories (which are basically space–time varying boundary conditions) may lead to the same current structural geological setting. Since the current geological setting is the only observation that can be made (one cannot go back in time unfortunately), much uncertainty remains in the initial / boundary or deformation histories.

Deterministic models are certainly useful as a good start to understand some of the processes taking place, to perform some initial investigation of what is possible. However, they have little "predictive" power, that is, they cannot be used to quantitatively forecast a phenomenon such as, for example, the oil production over the next few years or the travel time of a contaminant plume. A single model will yield a single value for the prediction, which will certainly differ from the truth because of the many sources of uncertainty.

The question is, therefore, not a matter of choosing deterministic models or not, but to state the question in terms of the purpose of the modeling exercise: what are we going to use these models for? What is the objective? Is the goal to simply have physical realism? Is it simply to perform a scientific study, do a few simulations, or to make actual quantitative predictions? In the latter case, an approach that includes the various sources of uncertainty, even when employing simpler models, may often be better. This theme of purpose-driven modeling will be argued throughout this book. In that sense, a deterministic model should not be excluded a priori.

3.4 Models of Uncertainty

The idea behind creating models of uncertainty is to tie the various sources of uncertainty mentioned above with deterministic (or not!) laws of physical and dynamic processes and make such modeling dependent on the purpose for which they are used. Figure 3.2 provides an overview that will be used throughout this book. Each of the following chapters treats various aspects displayed in this figure; therefore, each chapter will start by recalling this figure with the treated topics highlighted. This figure by no means represents the only view on modeling uncertainty, but it can be used to structure and solve many problems.

Since many Earth Science problems involve spatial modeling, we retain a specific box "spatial stochastic modeling" that describes the various techniques for creating a 3D (or 4D if time is considered) Earth model, such as shown in Figure 1.3. Typically this

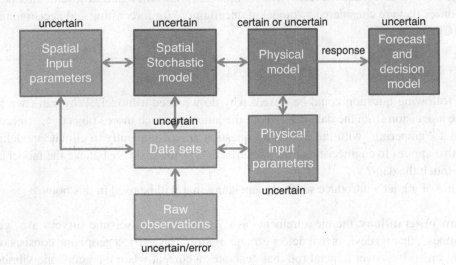

Figure 3.2 Overview of the various components of modeling uncertainty in the Earth Sciences. The arrows indicate the direction of interaction or relationship.

box includes many techniques covered in "geostatistics" or "spatial statistics," whereby various alternative property fields (often of a static property, such as a rock type or a surface structure, not for a dynamic property such as pressure) are created on a specified grid of cells through Monte Carlo simulation. Recall that in Chapter 2, Monte Carlo simulation of a single distribution function was explained. Chapters 6–8 describe techniques for extending this simple principle to more than just one variable at a time.

Spatial stochastic models require input parameters. Just like a Gaussian distribution (a 1D stochastic model in fact) requires two parameters, namely a mean and variance, a spatial stochastic model requires more complex "parameters". Spatial models and their corresponding input parameters are discussed in Chapter 5.

As discussed, physical models can be deterministic, for example a given Partial Differential Equation (PDE) that is solved numerically on a grid, or a simple law of mass conservation over a given volume of space. The physical processes themselves could be uncertain; in that case several alternative physical models could be defined. Input to the physical models are both static properties, for example, rock types, layering, faulting, or created from the spatial stochastic model as well as dynamic initial and boundary conditions (such as the initial pressure of a system, the stresses applied to the system), chemical (the initial compositions of fluids or gasses) and physical properties (ideal gas properties, compressibility) termed "physical input parameters" in Figure 3.1. Numerical modeling techniques (finite element or finite difference techniques) are then used to solve the partial differential equations representing the physical laws on the grid.

The outcome of the physical model, termed the response, is then used for forecasting an event, for example, change in temperature, or to make decisions, for example, reduce carbon dioxide in the atmosphere or take any other action affecting climate. In the subsequent chapters, we will create/generate several sets of input parameters, from which we create several spatial stochastic models as well as several alternative physical parameters (possibly also several physical models). This will result in several alternative responses that are considered a "model of uncertainty" for forecasting and decision making (Chapter 4).

3.5 Model and Data Relationship

The following question could be posed: why do we need a model? Why can't we just make a decision from the data itself? Isn't the latter approach more "objective," since we are not "tampering" with the data? This question arises frequently in climate modeling, but also applies to engineering of the subsurface: how much do we believe the model and how much the data?

First of all, let's introduce some nomenclature that will be used in this book.

- **Raw observations**: the measurement "as is". For example, seismic surveys use "geophones", that is, devices that detect ground motion. Basically, a geophone consists of a coil on springs over a metal rod that generates a current when the geophone vibrates; this current is then translated/filtered into the amplitude of the coil movement. So the raw observations are not the amount of vibration but the amount of current generated.

To obtain a seismic image, a lot of processing needs to done to take these raw observations from an array of geophones and create an image of the subsurface from it.

- **Data sets**: the processed and interpreted raw observations. A satellite provides only a "proxy" to temperature at the Earth surface; it does not measure temperature directly. Some filtering and modeling is required to turn the measurements taken by satellites into an actual data set of temperature that can be used by modelers. Seismic images or other geophysical images, such as shown in Figure 3.1, are data sets, not raw observations.

- **Information and knowledge**: any other information about the study in questions whether from experts, interpretations, analogues, previous studies, simulation models and so on.

Raw observations are therefore rarely of direct use in making decisions or in drawing conclusions about a particular study. Such observations often need to be "massaged," "processed" or "filtered" and such processes often require either some form of expert interpretation or a "model." The expert's brain doing this processing is in fact a type of human "model." Two different experts may create two different data sets from the same raw observations because each has a different "model" in their head. As a consequence, there is always an interaction or symbiosis between data sets and models; hence the arrows between models/parameters and data goes both ways in Figure 3.1. Theoretical, physical or mathematical models contain parameters or even heuristic rules that need to be constrained, derived or calibrated from data, while data sets are derived from raw observations through (physical) models. When multiple sets of different type of observations are collected (e.g., samples and geophysical data), then models are needed to "integrate" these data into a consistent theoretical framework, a topic that will be treated in Chapter 7. Data sets are therefore model dependent and models should be data driven. Hence, to address the above questions: there should not be a competition between data and model, rather models and data should be used in symbiosis to address inconsistencies in raw observations/data or model.

The third component in our nomenclature: information and knowledge is used as a placeholder for any ancillary information that is not derived directly from raw observations. Such knowledge, however, can be used to turn raw observations into data sets or to define theoretical models that apply.

3.6 Bayesian View on Uncertainty

In Chapter 2 a simple rule for determining a conditional probability of a random variable A, given the value outcome of another random variable $B = b$, was discussed:

$$P(A|B) = \frac{P(B|A)\, P(A)}{P(B)}$$

Although we considered A and B to be simple events, we can make these events as complex as we want, nothing in Bayes' rule prevents this. Consider as $\mathbf{A}$ the complete unknown Earth, or at least that area and the relevant properties that we would like to model, and $\mathbf{B}$ all the data available, possibly consisting of many data sources available $\mathbf{B} = (\mathbf{B}_1, \mathbf{B}_2, \ldots, \mathbf{B}_n)$, each $\mathbf{B}_i$ is a data set. Basically, both $\mathbf{A}$ and $\mathbf{B}$ are large vectors containing a large amount of variables and data to be considered. For example, $\mathbf{A}$ could contain all variables we want to model on a large grid, containing millions of grid cells, each grid cell representing one or multiple variable (such as a rock type). Nevertheless, Bayes' rule applies as well to these vectors

$$P(\mathbf{A}|\mathbf{B}) = P(\mathbf{A}|\mathbf{B}_1, \mathbf{B}_2, \ldots, \mathbf{B}_n) = \frac{P(\mathbf{B}|\mathbf{A})\,P(\mathbf{A})}{P(\mathbf{B})}$$

$$= \frac{P(\mathbf{B}_1, \mathbf{B}_2, \ldots, \mathbf{B}_n|\mathbf{A})\,P(\mathbf{A})}{P(\mathbf{B}_1, \mathbf{B}_2, \ldots, \mathbf{B}_n)} \simeq P(\mathbf{B}_1, \mathbf{B}_2, \ldots, \mathbf{B}_n|\mathbf{A})\,P(\mathbf{A})$$

Another way of viewing this is to define the model as $\mathbf{A}$, and the complete data as $\mathbf{B}$. Then, $P(\mathbf{A}|\mathbf{B}_1, \mathbf{B}_2, \ldots, \mathbf{B}_n)$ can be seen as a model of uncertainty of $\mathbf{A}$ given the data $\mathbf{B}$. This probability distribution describes the frequency of outcomes of $\mathbf{A}$ (if $\mathbf{A}$ contains only discrete variables) or the probability density if $\mathbf{A}$ contains continuous variables. Drawing samples from P by Monte Carlo simulation (done in Chapter 2 where A is a single variable and covered in Chapter 5 when $\mathbf{A}$ is multiple variables), would result in samples $\mathbf{a}_1, \mathbf{a}_2, \mathbf{a}_3, \ldots, \mathbf{a}_L$, if L samples are drawn. The set of samples, $\mathbf{a}_1, \mathbf{a}_2, \ldots, \mathbf{a}_L$ is also a model of uncertainty and would approximate P if L becomes infinite. However, luckily we need only a few hundred or thousand samples in most cases.

Bayes' rule states that there are rules in place to define a model of uncertainty $P(\mathbf{A}|\mathbf{B}_1, \mathbf{B}_2, \ldots, \mathbf{B}_n)$ or to create the samples drawn from it. In fact, there are two conditions implied by Bayes' rule:

1 It should depend on some prior model of uncertainty $P(\mathbf{A})$.

2 It should depend on the relationship between data $\mathbf{B}$ and model $\mathbf{A}$ as specified in $P(\mathbf{B}|\mathbf{A})$, termed the likelihood probability or simply "likelihood".

Consider first the prior uncertainty. Suppose we are studying a carbonate reef system in a subsurface formation. The prior outcomes of $\mathbf{A}$ are then defined as all possible reef systems that could occur in the world. So the possible outcomes of $\mathbf{A}$ could be very large. Now consider we have a specific data set of samples taken from the area, denoted as $\mathbf{b}$ (note the small letter used for an outcome) and a geologist has interpreted that this data set can only occur when the reef $\mathbf{A}$ being studied is a reef from the cretaceous era ($\mathbf{a} =$ "reef belonging to the cretaceous," small $\mathbf{a}$ for specific outcome), in other words, the data are fully indicative of a cretaceous reef, meaning that we can exclude reefs from all other geological eras, in other words:

$$P(B = b|A = a) = \begin{cases} 1 \text{ if a reef } A \text{ belongs to the cretaceous (or } A = a) \\ 0 \qquad\qquad\qquad\qquad\qquad\quad \text{else} \end{cases}$$

This likelihood need not always be 1 or 0, for example, it may have been determined by a geological expert (or via physical modeling) that the data sets **b** occurs with an 80% chance when the reef is cretaceous, then:

$$P(B = b|A = a) = \begin{cases} 0.8 \text{ if a reef } A \text{ belongs to the cretaceous (or } A = a) \\ 0.2 \qquad\qquad\qquad\qquad\qquad\quad \text{else} \end{cases}$$

In a true Bayesian sense the prior probability (or prior model of uncertainty) is determined *without* looking at any data. In reality, this is rarely the case. Modelers always look at data when performing a study. Indeed, why not look at possibly expensive data! By looking at data, it does makes sense to exclude what is completely impossible (such as all clastic systems when studying a carbonate system) and if the data are clearly indicative of this, then, "in spirit" with Bayes' rule, such prior exclusion could be done (although not without risk of excluding the unknown unknowns). Determining prior probabilities is difficult, highly subjective but may be critical in the absence of data. Prior probabilities can be determined from historical observations, if they are deemed good analogs for the studies situation (a considerable assumption). Another way to determine prior probabilities is to interview experts. This procedure is then termed prior uncertainty elicitation from experts, where elements of psychology are brought to the table to make the determination of probabilities as realistic as possible.

Bayes' rule suggests that at least a mental exercise should be made at collecting all possibilities imaginable prior to including the data into the model. This is not a bad idea and in line with the various sources of uncertainty discussed above, namely, one should try to imagine the unknown unknowns or the epistemological uncertainty (Albert Einstein's quote applies well here). Putting too much focus immediately on data is extremely tempting, since raw observations are the only "hard facts" listed on paper (or computers). Eliminating too many possibilities from the beginning may lead to creating an unrealistically small uncertainty. This is simply due to the fact that data are incomplete or prone to error (recall: we could have known/we roughly know).

The model of uncertainty $P(A|B_1, B_2, \ldots, B_n)$ in Bayesian terms is also called the posterior probability or posterior model of uncertainty (as in "after" considering the data). Bayes' rule states clearly the following: when specifying a prior uncertainty and the probabilistic relationship between data and model through a likelihood probability, then the posterior uncertainty is fully defined. In other words, there is no longer a degree of freedom to choose such posterior uncertainty without being in conflict either with the prior uncertainty and/or the likelihood. The more difficult term in practice is the prior uncertainty, since the likelihood can often be determined from physical models, as discussed in Chapter 7. In fact, any modeling of uncertainty is, particularly in cases with considerable uncertainty, only as good as the modeling of the prior, that is, the imagination of the modeler, referring to Einstein's quote. Various examples of this concept are described later in the book. In Chapters 5–8 various modeling techniques that use Bayes' rule in practice are also discussed.

This book will follow the philosophy enshrined in Bayes' rule and use it as a reference mathematical framework. One should understand that, given the subjectivity inherent to modeling uncertainty, probability theory and Bayes' rule need not be the only guideline or mathematical framework within which such modeling can be achieved. Other frameworks (such as, for example, fuzzy logic) are available but as of current have not proven to reach the same level of maturity in practice. It is, however, important to work within a given reference mathematical framework to make modeling uncertainty a repeatable, hence scientific, area of expertise.

3.7 Model Verification and Falsification

Is it possible to check whether a model of uncertainty is "correct"? Before addressing this question, consider a simpler question: can we verify/validate/check whether a single deterministic model is "correct"? To address this question, it is necessary to first state exactly what is meant by "validation" or "verification" and more importantly what is meant by "correct"!

A common way to validate a deterministic model is to check whether it either matches the data set or, in a more general sense, reproduces some patterns observed in Nature. For example, does our climate model match/reproduce the patterns observed in historic trends of temperate increase, carbon dioxide output and so on? Is this a good way to check such model?

Recall that deterministic models are often based on physical and dynamic processes. Such models are created based on inductive arguments, that is, a type of reasoning that attempts to draw conclusions (physical models) from facts (data, observation, information). For example, all swans in Europe are white (fact), therefore, there are only white swans (conclusion). True, until one discovered Black swans in Australia. Similar statements have been made about Newtonian physics, which was considered generally applicable until Einstein discovered relativity. No inductive proposition (such as a physical model) can be proven with certainty, even if it matches the data perfectly (only white swans are observed). Hence, if a deterministic model matches the data, then that does not verify (proof as truth) that model. It may, however, have been *validated* by the data, a less stringent term, meaning that there is some internal consistency in the model and no apparent flaws can be detected.

A famous philosopher, Karl Popper, went even further and stated that the induction approach to science, which emanated in the Renaissance era, should be repudiated; instead he argues for a methodology based on "falsification." According to his philosophies, physical processes are laws that are only abstract in nature and can never be proven, they can only be disproven (falsified) with facts or data. Note that the term falsifiable should not be mistaken for "being false": it means that if a scientific theory is false, then this can be shown by data or observations. Scientific knowledge is only as good as the creativity, imagination of the human brain. In line with our discussion on Bayes' rule, such reasoning can also be put in a Bayesian framework: the prior consists of all possibilities imagined by a human modeler, possibly aided by computers, then, the posterior includes only those possibilities that cannot be falsified with data as modeled in the likelihood.

Popper's view, even though it may be a bit too strict, can be very useful in many applications, since many times we are not interested in proving something through modeling but in disproving certain theories or scenarios. Knowing that something cannot occur (such as an earthquake) may have important societal consequences.

We return to our broader question if it is possible to check whether a model of uncertainty is "correct." Such correctness would call for a definition of what is correct: does it match data? Does it represent physics well? But in view of Popper's philosophy such reasoning is flawed even if a definition of correctness can be defined (which is subjective to start with), it still cannot be proven to be true, it can only be proven to be false. This is certainly the case for models of uncertainty, which can in fact never be objectively proven true neither false, since the statement, "the chance of finding gold is 60%," cannot be verified as false or true objectively. Gold will either be found or not, hence, even in hindsight, this statement cannot be disproven.

3.8 Model Complexity

An important question to ask in any modeling, including any modeling of uncertainty, is how simple or complex a model should be. Complexity could entail creating models with more grid cells or more variables that can resolve better fine scale features or by including more complex physics or more complex spatial models containing more input parameters. A principle that is often quoted in addressing this question is "Occam's razor," named after a fourteenth century logician: "entities must not be multiplied beyond necessity," under which is often understood that, in a modeling context, when competing models are equal in various respects, the principle recommends selection of the model that introduces the fewest parameters/variables and simpler physics while at the same time equally sufficiently answering the question. In other words, a simple model is preferred over a complex one, if both yield the same result. In Popper's view, simple models are also preferred because they are more easily falsifiable. Note that Occam's razor is a principle not a law, it may guide modelers; it is, however, not considered as being scientifically correct nor true. In this book, we will start from this principle but add that no model should be more complex than needed for a given purpose or objective. Hence, it is the aim of modeling that should drive model complexity, not necessarily some principle of parsimony. If the objective of modeling changes then model complexity should change as well. Figuring out how complex models need to be for a given purpose or decision process is not trivial; some techniques are discussed in Chapter 10.

One can now start to understand that there is a trade-off between an increase in model complexity and an increased importance of prior information on models and modeling parameters. Modeling a phenomenon with simple physics and few variables may give a false sense of security if the modeling parameters can be deterministically determined (for example by inverse modeling; Chapter 7) from the data: the model matches the data and only one parameter set can be found that satisfies this condition. However, often, when more data become available, a mismatch may be observed between the deterministic model and the new data. To address this problem, one often opts for more complex modeling in order to match the new data in addition to the existing data in the belief that the problem lies in not having the correct physical model in the first place. However, increasing the complexity of the physical model requires more parameters (for example going from a 1D model to a 3D model) that cannot be deterministically determined from data, that is, uncertainty in the model increases, which seems a bit strange since more data are obtained! Bayes' rule then states that such uncertainty becomes increasingly function of the prior uncertainty of the modeling parameters.

3.9 Talking about Uncertainty

Using a proper language is often critical to conveying the right message. As one starts to understand from this chapter that modeling uncertainty is full of potential pitfalls and misconception, we need to start "talking about uncertainty" using a wordage that is in line with the philosophical questions and concepts raised. I advocate the following wordage:

- quantifying uncertainty

- assessing uncertainty

- modeling uncertainty

- realistic assessment of uncertainty.

   While I consider as confusing at best and possibly incorrect use of wordage as:

- estimating uncertainty

- best uncertainty estimate

- optimal uncertainty

- correct uncertainty.

   The latter terms are confusing, because words such as "estimation," "best," "correct" or "optimal" call for the definition of what is best or optimal (note that any estimation in a rigorous statistical sense requires a loss function, but we won't go into detail here). Hence such definition requires imposing a criterion of difference with the actual true value or truth, which as discussed above is not possible in the case of modeling uncertainty. There is no "true probability." Unfortunately, these confusing terms are pervasive in the scientific literature even in journals such as Science and Nature.

3.10 Examples

In this section the various concepts and philosophies are illustrated with some examples of modeling in the Earth Sciences. The intent is not to go into detail of the modeling itself; techniques are discussed in subsequent chapters. The aim is to illustrate the various concept of uncertainty, the issues of scale, deterministic vs stochastic modeling and the subjectivity of models of uncertainty.

3.10.1 Climate Modeling

3.10.1.1 Description

Climate models are computer simulations based on physical laws of atmospheric conditions, usually simulated over long periods. Early models were based on principles of energy balance (no grid required) aiming to compute the average temperature of the Earth from contributions such as solar radiation and the atmospheric gas concentrations. This "zero dimensional" (the Earth is basically a point mass) model was then extended to one and two dimensional models where temperature is function of latitude and longitude. Current models are three dimensional and termed (atmospheric) general circulation models (AGCM or GCM). OGCM are then oceanic general circulation models.

These models contain usually a few million grid cells covering the Earth atmosphere, each grid cell is on average 100 km (order of magnitude) in latitude and 250 km in longitude and has roughly 20 vertical layers, depending on the kind of modeling used. Each grid cell typically contains four variables (windspeed (2), temperature and humidity). Equations of state (thermodynamics) compute the various effects such as radiation and convection within a grid cell, while equations of motion calculate fluxes into the neighboring grid cells. Evidently, such models vary in time, with time steps in the order of 10 minutes. Since simulations are done over a century, the CPU demand is large and typically 10–100 hours of computing time depending on the model complexity.

Climate models are coarse, certainly compared to numerical weather prediction models which aim to forecast weather on a regional scale. The coarseness of the model raises concerns since many small scale processes, such as the movement of clouds (1 km scale), may have a considerable impact on climate. To account for such subgrid scale processes, climate modelers use what is called "parameterization." Clouds are not represented as they really exists (as a convection column for example), rather the amount of cloud cover within a grid cell is calculated as some function of temperature and humidity within that grid cell, so the clouds' formation is not modeled explicitly but parameterized by other variables. In doing so, there is an assumption that small scale processes (clouds) can be represented by large scale variables (temperature).

3.10.1.2 Creating Data Sets Using Models

Climate models need initial conditions and boundary conditions, such as for example the sea surface temperature. Many kinds of instruments and measurements taken under many kinds of conditions (arctic vs equatorial) are needed to initialize the three dimensional grid. The quality of instruments changes over history, hence a measurement taken

50 years ago should not be taken on the same footing as a measurement taken today. It is known that satellites provide only proxy measurements of temperature possibly distorted by optical effects. This means that any "raw observations" related to temperature require a lot of filtering, processing and interpolation (all "models") before a data set can be created. Hence "models" are needed to create data sets from raw observations.

3.10.1.3 Parameterization of Subgrid Variability

Parameterization, as discussed, is a highly simplified way of representing small scale variations of cloud formation on a coarse grid. An assumption is made that small scale variation can be explained locally by a large scale variable (such as temperature). This is a major physical assumption, since it may be that small scale cloud formation has a more regional impact than a simple local impact (grid-block scale), hence such a model is not necessarily a "correct" representation of what is really happening. Climate modelers therefore need to "tune," that is, to adjust their parameterization to match the observed data. Such *ad hoc* tuning is also evidence of the lack of understanding (i.e., uncertainty!) of the impact of clouds on climate models. While such tuning may lead to matching the data better, it is not a guarantee of predicting future behavior better if such tuning is simple, an *ad hoc* trick and does not represent an actual physical phenomenon.

3.10.1.4 Model Complexity

How complex or simple should climate models be and how should we judge the level of complexity? It seems often the case that physically more realistic models are preferred over simpler models: greater complexity equals greater realism. However, greater complexity also leads to greater CPU times resulting in the ability to run only a few models if each model run takes hundreds of hours. This means that exploring possibilities, that is, uncertainty, becomes impossible and models are merely deterministic. Yet, the question should be asked whether such complex modeling is really needed for the purpose presented; a model is then only "better" with respect to the objective stated, not just because of increased physical realism, which is a very specific objective. Such an objective could be to predict the mean temperature increase, the increase in carbon dioxide concentration in the atmosphere or on regional forecast of climate change. Each such purpose may lead to a different model complexity. If simpler models can be run more frequently when uncertainty is a critical objective (such as in forecasts), then simpler models may be preferred if the difference between a simple model and a complex model is small compared to all other uncertainties in the model. Indeed, why care about making a small error in taking a simpler model as compared to a complex model when such error disappears compared to all other sources of uncertainty.

3.10.2 Reservoir Modeling

3.10.2.1 Description

A reservoir is a term used for a porous media containing fluids and/or gasses (liquids). A reservoir is often considered as closed, that is, the liquids are trapped, but in a broader sense it may also be considered as open, such as is the case for aquifer systems. Reservoir modeling is a broad term used for describing the 3D modeling of the porous media and the flow of fluids and gasses in this medium due to some external "stress," such as a pumping or injection well. Most common examples of reservoirs are petroleum and gas reservoirs, but reservoirs are also used for storing gas or for sequestering carbon dioxide. Reservoir engineering is used as a term for the design of the extraction or injection of liquids and such engineering often makes use of reservoir models to predict or forecast the reservoir behavior (flow of liquids) under such change. This allows engineers to optimize the system, for example by maximizing production. Reservoir modeling requires the confluence of many sciences such as geology, geophysics, physics and chemistry. In fact, it has many similarities in terms of modeling uncertainty as climate modeling.

Fine scale (or high resolution) reservoir models contain often tens of millions of cells, each cell can be as small as 0.3 m in the vertical and 15 m in the horizontal depending on the size of the reservoir. Many kinds of data sets are available: wells are drilled and cored, also logged (a sounding instrument), geophysical surveys are taken (such as TEM, seismic) to produce some sort of 3D image of the subsurface geology, wells can be tested by briefly injecting fluids into or producing fluids in the well and observing the pressure response. The production of fluids is often simulated using a numerical simulation of flow in porous media. Such simulations are very CPU demanding, particularly if complex physics is included, such as changes in phase behavior due to increasing/decreasing pressure or gravity effect (fluids that move downward). Such "flow simulations" are infeasible on a 10 million cell grid, hence fine scale models are coarsened to a model containing only 100 000 cells. Even then a flow simulation may take 10 hours of CPU time.

3.10.2.2 Creating Data Sets Using Models

Reservoir models use various data sources, such as geophysical data (e.g., seismic data) and data obtained through "logging" the well-bore. Logging tools are basically also "geophysical" devices, they emit a certain source wave, for example, a pressure wave, gamma wave or neutron beam, into the geological formation and measure how the formation responds. Well-log modeling and interpretation then consist of turning these signals into interpretations of either sand content, water content, along the well bore depending on the type of signal used. Turning "signals" into an actual vertical succession of sand/shale requires a large degree of physical modeling as well as interpretation by an expert; hence any data set of sand/shale sequence in a well bore is itself largely a model interpretation from the raw observations obtained from this tool.

3.10.2.3 Parameterization of Subgrid Variability

Porous material often consists of grains, cement holding the grains together and pores containing the liquids. Reservoir models can, however, not represent accurately every grain in the subsurface, yet flow is dependent on the arrangement and packing of the grains. Similar to climate models, the fine scale variability at the pore level needs to be "parameterized", that is, some models needs to represent the average behavior of flow taking place in the pores at the grid cell level (dimension tens of feet). If multiple fluids are present in the pores, such as oil and water, or the arrangement of grains is in complex layering (called stratification), then such parameterization can be rather complex, dependent on the geometry of the arrangement and the proportion of oil vs water in the system (the term relative permeability curves is used in reservoir engineering for such parameterization). As in climate modeling this parameterization requires tuning, either from cores extracted from the reservoir or from production itself, since during production one measures water and oil rates that are dependent on the grain size distribution.

3.10.2.4 Model Complexity

Reservoir modeling is never a goal on its own. Models are built for engineering purposes and for making decisions about if we should or how we should produce the reservoir. It makes sense to build the models at the level of complexity (either geological or flow physics complexity) as is required to solve the particular engineering question. However, in practice it is often the data and the modelers' preference that drives the model complexity. Often deterministic models representing as much as possible the physical realism as well as the data observed are chosen. An important data source in this regard is historic production from the reservoir, just as the historic temperature variation is an important source to calibrate climate models. The "quality" of the model is often judged by how well these data are matched, and the models are built at a complexity that can match that data. However, decisions made on the reservoir may depend on reservoir features that are at a scale much smaller than required to include accurately the physics and the data. Therefore, the current trend is toward purpose driven and stochastic modeling with a decision analysis framework. Several chapters in this book illustrate this succinctly.

Further Reading

Bardossy, G. and Fodor, J. (2004) *Evaluation of Uncertainties and Risks in Geology*, Springer Verlag.

Caers, J. (2005) *Petroleum Geostatistics*, Society of Petroleum Engineers, Austin, TX.

Shackley, S., Young, P., Parkinson, S., and Wynne, B. (1998) Uncertainty, complexity and concepts of good science in climate modeling: are GCMs the best tools? *Climatic Change*, **38**(2), 159–205.

Taleb, N.N. (2010) *The Black Swan*, Random House USA, Inc.

Popper, K. (1959) *The logic of scientific discovery*, Routledge Publishers, 1959.

4

Engineering the Earth: Making Decisions Under Uncertainty

A decision can be defined as a conscious, irrevocable allocation of resources to achieve desired objectives. Building realistic models of uncertainty, in the context of decision making, is what this book is about. Decision making and uncertainty modeling are integral and synergetic processes, not a sequential set of steps. Certainly, in engineering applications, no model of uncertainty is relevant or even useful without a decision goal in mind.

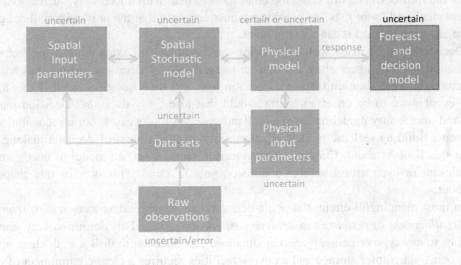

4.1 Introduction

Making good decisions is important in many aspects of life. Decisions in the personal realm are made by individuals and usually consider the consequences of those decisions on others (e.g., family members). In organizations (e.g., corporations, governments, universities, etc.), individuals also play a critical role in decision making but are usually part

Modeling Uncertainty in the Earth Sciences, First Edition. Jef Caers.
© 2011 John Wiley & Sons, Ltd. Published 2011 by John Wiley & Sons, Ltd.

of a group-based decision making process. How does an individual or an organization know whether they are making a good decision at the time they are making that decision (without the benefit of hindsight)? The question therefore begs: "Would you know a good decision if you saw one?" Well, it depends. Without any field specific knowledge one could be inclined to define decision making as "choosing between many alternatives that best fit your goals." However, the evident questions then are (1) how to define what is best or optimal, one needs some criterion and the decision may change if this criterion changes, and (2) what are your stated goals? Decision analysis theory provides sound scientific tools for addressing these questions in a structured, repeatable way.

Uncertainty has an important role in making sound decisions. The existence of uncertainty does not preclude one from making a decision. In fact, decision making under uncertainty is the norm for most decisions of consequence. For example, in the personal realm you already have experienced decision making under uncertainty when you went through the college application process. Most readers of this book probably applied to several colleges because there was uncertainty on gaining admission to the college of one's choice. Similarly, universities make offers to more students than they can accommodate because there is uncertainty on how many students will accept their offer. Universities often rely on historical data on acceptance rates to decide how many offers to make. However, there is uncertainty because the past is not usually a "perfect" predictor of the future. Universities develop other tools to deal with uncertainty, such as "early acceptance" and "wait lists," to provide a higher certainty that the university will not get more acceptances than it can accommodate.

Decisions can be made without knowing the hard facts, an exact number, perfect information. In fact, uncertainty is often an integral part of decision making, not some afterthought. One shouldn't make a decision first and then question, what if this and that event were to be uncertain? How would that affect my decision? Decision making and uncertainty modeling are integral and synergetic processes, not a sequential set of steps. Building realistic models of uncertainty, in the context of decision making is what this book is about. Certainly in engineering application, no model of uncertainty is relevant or even useful without a decision goal in mind. This is what this chapter is about.

In most meaningful circumstances, a decision can be defined as a *conscious, irrevocable allocation of resources to achieve desired objectives*. This definition very much applies to any type of geo-engineering situation. The decision to drill a well, clean up a site, construct aquifer storage and recovery facilities requires a clear commitment of resources. One may go even to a higher level and consider policy making by government or organizations as designed to affect decisions to achieve a certain objective. For example, energy legislation that implements a tax on carbon can impact commitment of resources as related to energy supply, energy conservation, carbon capture technology solutions. The 10-point program in Figure 1.1 outlines clear objectives that will require allocation of resources to make this happen.

At Stanford University, Professor Ron Howard has been a leader in the field of decision analysis since 1966. He described this field as a "systematic procedure for transforming opaque decision problems into transparent decision problems by a sequence of transparent steps." Applying the field of decision analysis to the Earth Sciences is not trivial. Several challenges need to be overcome:

- **Uncertainty**: while most of this book deals with the Earth aspect of uncertainty as pertains to the measurements and models established to make prediction and optimize profit or use of resources, there may be many other sources of uncertainty, more related to the economic portion of uncertainty (costs, prices, human resources), that are not discussed in this book.

- **Complexity**: rarely do we make a single decision on a single decision question. Often a complex sequence of decisions needs to be made. For example in developing reservoirs, all wells are not drilled at one point in time. Rather wells are drilled sequentially. New data become available after drilling each well that impact the picture of uncertainty and subsequent decisions of allocating resources (e.g., future drilling).

- **Multiple objectives**: often, there are competing objectives in decision making, for example as related to safety and environmental concern compared to the need for energy resources. Any climate policy or decisions made around this issue will likely involve many, possibly conflicting objectives.

- **Time component**: if it takes too much time to build a model of uncertainty that tries to include all sorts of complexity, and the decision must be made in a much shorter time frame, then a complex model ends up having little input into the decision. This is often the case in time-sensitive businesses or industries (competitive or oil field reserve calculations, for example). In such cases, one may want to opt for simpler models of uncertainty over complex ones.

This book provides a brief introduction to decision analysis emphasizing the most important elements for modeling uncertainty in the Earth Sciences. Many ideas in writing this chapter were borrowed from the recent primer publication "Making good decisions" by Reidar Bratvold and Steve Begg.

4.2 Making Decisions

4.2.1 Example Problem

Consider an area, as shown in Figure 4.1, where a point source of pollution due to leakage of chemicals was discovered in the subsurface. This pollution source is close to a well used to supply drinking water. While it hasn't happened yet, some speculate that this pollution may travel to the drinking well due to the geological nature of the subsurface. The study area consists of porous unconsolidated sand in a nonporous clay material. The deposit lies over an impermeable igneous rock. Some basic geological studies have revealed that this is an alluvial deposit. Some geologists argue that the main feature of these deposits is channel ribbons such as in Figure 4.1. Only very basic information is

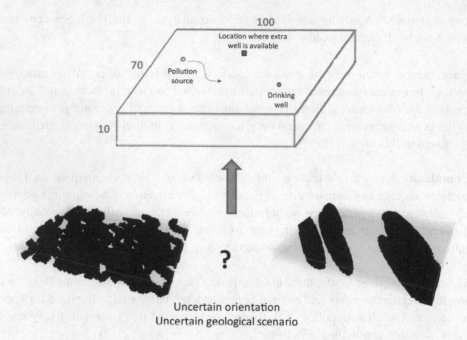

Figure 4.1 Example of a contaminant that may reach a drinking well, dependent on the sub-surface geological heterogeneity.

known about this type of channel based on analog information and an outcrop in a nearby area, but most important is the channel orientation Θ_{ch}, which may impact the direction of the contaminant flow; in fact two possibilities are assessed:

$$P(\Theta_{ch} = 150°) = 0.4 \quad P(\Theta_{ch} = 50°) = 0.6$$

However, geologists of equal standing claim that this area does not contain sand chan-nels but sand bars (Half-Ellipsoid) that tend to be smaller than long sinuous sand chan-nels. Similarly, the orientation of these elliptical bars is important and two possibilities are assessed:

$$P(\Theta_{bar} = 150°) = 0.4 \quad P(\Theta_{bar} = 50°) = 0.6$$

Some believe that there are enough barriers in the subsurface to prevent such pollution and claim that the pollution will remain isolated and that clean-up would not be required. Some believe that even if the pollutant reaches the drinking well the concentration levels will be so low because of mixing and dilution in the aquifer that it will not pose a health concern, and hence investment in clean-up is not required.

The local government has to make a decision in this case: either to act and start a clean-up operation (which is costly for tax-payers) or do nothing, thereby avoiding the clean-up cost but potentially be sued later in court (by local residents) for negligence when drink-ing water is actually contaminated. What decision would the local government make?

Clean-up or not? How would it reach such decision? Are there any other alternatives? For example, monitoring the drinking water at the surface and cleaning up the produced water at the surface if contamination is detected or importing "clean" water from another source? This is the subject of decision analysis. In this book the basic concepts that are relevant to solve problems such as the one described in this case are reviewed.

4.2.2 The Language of Decision Making

The ultimate goal of decision making is to generate good outcomes. However, good decisions should not be confused with good outcomes. If there is no uncertainty about future events, such as channel orientation in Figure 4.1, then a good decision will necessarily lead to a good outcome, assuming a rational decision maker will always choose the best possible outcome. However, in the presence of uncertainty, a bad outcome does not necessarily mean that a bad decision has been made. In other words, in the presence of uncertainty, a good outcome cannot be guaranteed. Evaluation of the decision making process should always take place over the long run, where, if the process promotes good decisions consistently, then the outcomes are likely to be consistently better than if no optimal decisions had been made. Unfortunately, a common mistake in many decisions or other scientific analysis that pertains to uncertainty is that such reasoning over the long run is often neglected and specific decisions are judged in the moment.

The first step in any decision making process is to structure the problem and to identify the main "elements." This by itself is the subject matter of a great deal of research and many books, some even including elements of human psychology and sociology. Often the structuring is the most important part of decision making, but that is not the main focus of this book.

The following five elements have typically been identified as part of the decision structuring process; each element carries specific definitions: (1) Alternatives, (2) Objectives (3) Information or knowledge/data sets (Chapter 3) (4) Payoff of each alternative for each objective (5) Decision (the ultimate choice among the alternatives).

Decision: a conscious, irrevocable allocation of resources to achieve desired objectives. A good decision is, therefore, an action we take that is logically consistent with the objectives stated, the alternatives we believe there to be, the knowledge/information and data sets we have and the preferences we have. Decision making is possible if there were no (mutually exclusive) *alternatives* or choices to be decided on. Alternatives can range from the simple yes/no (e.g., clean up or not), through the complex and sequential (e.g., oil and gas exploration, field development), to those with extremely large numbers of alternatives.

Rational decision making requires clear *objectives* that will be used to compare each alternative. This requires some measure (quantitative) to determine how well each alternative achieves the stated objective. Such measure is often termed an *attribute*. Because there may be multiple objectives, each carrying a different weight, some statement of preference for each objective in a multi-objective decision framework is required. Identification of objectives is driven by high level *values* of the decision maker.

A *payoff* is what finally happens with respect to an objective, as measured on its attribute scale, after all decisions have been made and all outcomes of uncertain events have been resolved. Payoffs may not be known exactly because of uncertainty and need to be predicted.

4.2.3 Structuring the Decision

The goal of this phase is to identify and establish the relationship between the various elements of making a decision. It is probably the most consequential part of decision making, since all the rest depends on it. It is also the least amenable to quantitative analysis and relies on the creativity of those working on making a decision. An important aspect of that creativity is to identify all realistic alternatives. Leaving out realistic alternatives has been identified as a fatal flaw, in hindsight, in important decisions. In some cases, this phase may enable a decision maker to make a decision with a clear understanding of alternatives because defining the decision in a structured context provides insights that were not apparent beforehand; a full quantitative analysis may not be needed. Also, even if a full quantitative analysis of the decision problem is completed, the numbers resulting from this analysis need not necessarily be taken at face value, they should be used as indicative instead of exact (recall the discussion on uncertainty), or be used for further sensitivity analysis: what really impacted our decision and can we do something about it?

Important to making a decision is to define the decision context, that is, the setting in which the decision occurs. Note that the same decision problem may occur in different contexts. The context will identify relevant alternatives and set the objectives. The context will also identify the decision maker, that is, that person whose objectives and preferences are required. In the context, the necessary assumptions and constraints need to be identified as well.

Once the context is defined, a set of objectives can be generated and their associated attribute scale with which to measure the value created by different decision alternatives identified. As shown in Figure 4.2, setting objectives is achieved through a value tree. This tree is generally developed by working from high level values to specific

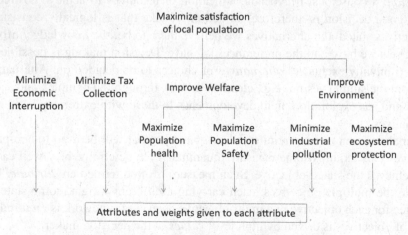

Figure 4.2   Value tree and hierarchy.

objectives. "Values" are general in nature; for example, values could be "be popular," "support UNICEF," "be healthy," "make money," while objectives are specific and could be of the form "maximize this" or "minimize that." Distinguish between *fundamental objectives* that identify the basic reasons why a decision is important and *means objectives*, which are ways of achieving a fundamental objective. Fundamental objectives should be independent and can be organized in a hierarchy. For example, "maximize profit" can be divided into "minimize cost" and "maximize revenue". Means objectives are not the fundamental reason for making a decision, a means objective could be, for example, to "create a clean environment" or to "have welfare programs". Indeed, welfare programs and a clean environment are only a means to population happiness.

The tree in Figure 4.2 could be used as a template for the local government in our example case.

The next step is to measure the achievement of an objective. For certain objectives, there will be a natural scale, either in dollars or ppm or rates. For other, more descriptive objectives a scale needs to be constructed, usually through numerical or other "levels" (high, medium, low). In our example of Figure 4.2 some objective, such as "Minimize Tax" has a natural scale in dollars, while others need a constructed scale. For example, "Maximize population safety" can be measured as

1 = no safety

2 = some safety but existing violent crime such as homicide

3 = no violent crime but excessive theft and burglary

4 = minor petty theft and vandalism

5 = no crime.

4.2.4 Modeling the Decision

The goal of this phase is to reach a preliminary decision based upon the alternatives that have been identified, the objectives that have been set, and the preferences for the relative importance of those objectives. Generally three steps are identified at this stage:

1 **Estimate payoff**: Make an assessment of the extent to which each alternative helps to achieve each objective (its payoff). The goal of this step is to make a relative comparison of the specific (not high level as discussed before but actual) value of the alternatives toward achieving objectives. This is done through the development of a payoff or consequence matrix (that quantifies how well each alternative score on the objective attribute scales). The second is to determine how much value is derived from these scores.

2 **Evaluate preference**: Determine the relative priority or preference for the objective. "Preference" in this context is used to describe the relative desirability between different objectives.

3 **Combine** the performance against each objective into an overall score for each alternative.

4.2.4.1 Payoffs and Value Functions

Recall that a payoff is the extent to which an objective is met after the decision is made and the outcomes of any uncertain events have been resolved. Therefore, payoffs are not known in advance and must be forecast or estimated. This may require a substantial amount of modeling and many such modeling techniques are covered in subsequent chapters.

Consider the value tree of Figure 4.2 and the decision alternatives whether to "clean up" or "not clean up". Assume that neither alternative will impact safety (cleaning up or not will not affect crime), then Figure 4.3 could be an example of a payoff matrix for this case. Tax collection will be impacted by such clean up because of its cost (say $10 million), which affects the local budget; however, ecosystem protection will increase (a constructed scale) while industrial pollution (in ppm) will be small (some pollutants may be left in the ground). In the case of not cleaning up, the tax collection also increases because of the requirement to import "clean" water to meet the needs of the population, assuming that the government would have to pay for this (suppose the contamination was made by a government research laboratory for example). This number is a little more difficult to establish. Indeed, the lawsuit will occur when the geology is unfavorable, causing the pollution to leak to the drinking well. But since the subsurface is unknown (it is not known whether there are channels or not, and what the orientation is), the number listed here is an expected payoff. To obtain this number, it is necessary to do some 3D modeling of the subsurface geology as well as simulate the flow of contaminants in the subsurface and figure out the chances of contamination from these simulations. These are topics treated in the rest of the book, so assume for now, that this expected payoff has been established.

	alternatives	
Objectives	Clean up	Do not clean up
Tax collection (million $)	10	18
Industrial pollution (ppm/area)	30	500
Ecosystem protection (1–5)	4	1
Population health (1–5)	5	2
Economic interruption (days)	365	0

Figure 4.3   Example of a payoff matrix.

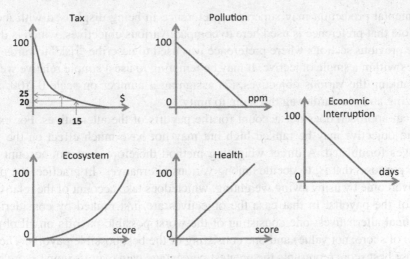

Figure 4.4 Example of value functions.

In a payoff matrix, it makes sense to only include objectives that distinguish among alternatives. Any other objective should be removed, such as population safety in this case. Also, in a payoff matrix, one works across the rows of the payoff matrix rather than down its columns.

The next evident question is how to incorporate our *preference* into a single attribute scale and combine payoffs measured on different scales. This is addressed by means of value functions. Value functions transform attributes to a common scale, say from 0 to 100. The value function expresses how an increase in the score translates into an increase in value. A linear value function (Figure 4.4) therefore states that such increase is proportional, such as for health, or inversely proportional, such as for pollution. A nonlinear function, such as for taxes in Figure 4.4, states that an increase in the amount collected results in a smaller decrease in actual value (high value if less taxes are collected). This means that if tax becomes larger, then any increase in tax will leave the population not necessarily equally more displeased (low value); they are already displeased with such high taxes! For the ecosystem, an opposite attitude could be argued for: more pollution will eventually completely ruin the ecosystem, while a small increase can possibly be tolerable. Such nonlinearity in the function can therefore be interpreted as the attitude towards "risk" that one may have about certain outcomes. For example, the attitude toward safety may be different than the attitude toward income. The preference may be to risk more when money is involved (tax) than with the environment, as such effects are often irrevocable (although governmental attitudes around the world may substantially vary in this aspect).

4.2.4.2 Weighting

Different objectives may carry different weights. This allows the decision maker to inject his/her preferences of one objective over another. For example, preference in environmental protection may supersede preference in being displeased with increased taxes. Note that preference is used here to compare various objectives, which is different from the previous sections where preference was used to describe "risk" towards various outcomes within a single objective. It may be tempting to use a simple relative weighting by (1) ranking the various objectives, (2) assigning a number on scale 0–100, and (3) normalizing and standardizong the score to unity.

Such an approach does not account for the payoffs of the alternatives. For example, a specific objective may be ranked high but may not have much effect on the various alternatives formulated. A direct weighting method therefore does not account for the ultimate purpose, that is, to decide among various alternatives. In practice, the problem can be overcome by using swing weighting, which does take account of the relative magnitudes of the payoffs. In that case the objectives are first ranked by considering two hypothetical alternatives: one consisting of the worst possible payoffs on all objectives (in terms of score, not value) and one consisting of the best possible payoffs. The objective whose best score represents the greatest percentage gain over its worst score is given the highest rank, and the methodology is repeated for the remaining objectives until all are ranked.

Since in our example case we are dealing with a binary decision, the weighting problem does not present itself (there is always a best and worst). To illustrate the swing weighting, consider a slightly modified example where two more alternatives are added: (1) a detailed clean up that is more costly but removes more contaminant, therefore protecting health and environment; and (2) to clean up half, that is, leave some pollutant behind with a decreased risk, but nonetheless risk of drinking water contamination. Figure 4.5 shows how swing weighting works. Firstly, the best and worst scores for each objective are taken, then the relative differences are ranked, with one being the largest relative difference. Clearly the tax impact is least discriminating among the alternative and therefore gets the highest rank (and smallest weight, as shown in Figure 4.6).

| | alternatives | | | | | | |
	Detailed clean up	Clean up	Partial clean up	Do not clean up	Best	Worst	Swing rank
Tax collection (million $)	12	10	8	18	8	18	5
Industrial pollution (ppm/area)	25	30	200	500	25	500	2
Ecosystem protection (1–5)	5	4	2	1	5	1	3
Population health (1–5)	5	5	2	2	5	2	4
Economic interruption (days)	500	365	50	0	500	0	1

Figure 4.5   Example of swing weighting.

Objectives	rank	weight	Detailed clean up	Clean up	Partial clean up	Do not clean up
Tax collection	5	0.07	30	20	100	0
Industrial pollution	2	0.27	100	99	40	0
Ecosystem protection	3	0.20	100	75	25	0
Population health	4	0.13	100	100	0	0
Economic interruption	1	0.33	0	33	90	100
		Total:	62.1	67.0	52.5	33.0

Figure 4.6 Scoring each alternative.

After weights and attributes are known, scores on each objective can be combined to determine an overall value for each alternative. This is achieved by calculating the weighted sum of each column in the value payoff matrix

$$v_j = \sum_{i=1}^{N_j} w_i v_{ij}$$

with w_i the weights calculated for each objective and v_{ij} the payoff of the j-th alternative for the i-th objective. This is done in Figure 4.6 where attributes are now turned into values using some arbitrary value functions (not shown). Therefore, in summary, the clean up alternative is the one that is logically consistent with maximizing the value of the decision, for given alternatives, objectives, weights, payoff predictions and preferences expressed in value functions.

4.2.4.3 Trade-Offs

Conflicting objectives can make decisions hard. In our case the minimization of tax burden is opposite the cost of maintaining a clean environment. Increasing returns (money) may come at the expense of increasing risks (health, safety, environment). A term called "the efficient frontier" may help us investigate what kind of trade-offs we are making and possibly change our decision based on this insight. This is very common in portfolio management (choice of equities (shares in stocks of companies) and bonds). Portfolio management utilizes historical data on return of equities to form the basis for assessment or risk and return and use the past performance as a proxy for future performance.

To study trade-offs, two categories are made: one for the risks and one for the returns (or cost/benefit). Overall weighted scores are then calculated for each subset, in a similar fashion to above, as given

$$v_j^{risk} = \sum_{i=1}^{N_{risk}} w_i v_{ij} \qquad v_j^{return} = \sum_{k=1}^{N_{return}} w_k v_{kj}$$

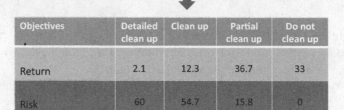

Objectives	rank	weight	Detailed clean up	Clean up	Partial clean up	Do not clean up
Tax collection	5	0.07	30	20	100	0
Economic interruption	1	0.33	0	33	90	100
Industrial pollution	2	0.27	100	99	40	0
Ecosystem protection	3	0.20	100	75	25	0
Population health	4	0.13	100	100	0	0

Objectives	Detailed clean up	Clean up	Partial clean up	Do not clean up
Return	2.1	12.3	36.7	33
Risk	60	54.7	15.8	0

Figure 4.7   Obtaining trade-offs from a payoff matrix.

where N_{risk} is the number of objectives classified as "risk" and N_{return} the number classified as "return" (Figure 4.7).

It is possible to plot the risk/return or cost/benefit in a plot such as Figure 4.8. From this plot some obvious alternatives can be eliminated as follows. Consider the alternative "partial clean up." The alternative "do not clean" up is clearly dominated by the alternative "partial clean up." Indeed, "partial clean up" has both more return and less risk. Therefore, the alternative "do not clean up" can be eliminated because it lies below the efficient frontier and results in taking on more risk relative to the return. "Do not clean up" is the only alternative that can be eliminated as such; other alternatives trade-off either risk or return with each other. The curve connecting these points is the efficient frontier. The efficient frontier can be seen as the best set of trade-offs between risk and return for the current alternatives. Recall that a decision can only be as good as the alternatives formulated. Therefore, pushing the efficient frontier upwards (i.e., up and towards the right in Figure 4.8) would require different alternatives leading to a better set of trade-offs. Such alternatives are only as good as the imagination of those creating them.

Figure 4.8 allows the question to be asked "Am I willing to trade-off more risk for more return between any two alternatives." For example, is the decrease of about five units of risk, worth the decrease in about ten units of return when going from "clean up" to "detailed clean-up"? If all attributes were in dollar values than these would be actual dollar trade-offs, in our case these are only indicative trade-offs, basically forming a scale from "less preferred" to "more preferred" in terms of trade-off. Note that in this discussion it is assumed that the scores are deterministic quantities (or some representative average value). Defining dominance is not quite so straightforward because of uncertainty. In

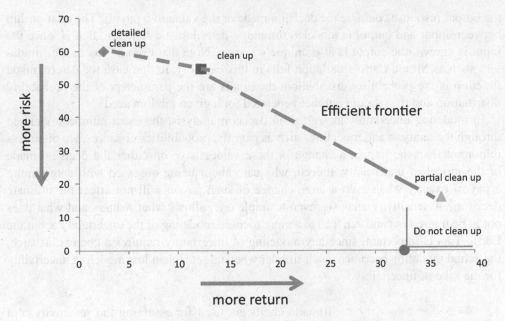

Figure 4.8 Assessing trade-offs between conflicting objectives.

this case, *stochastic* (or probabilistic) *dominance* can be used; this is discussed in more advanced books.

4.2.4.4 Sensitivity Analysis

Sensitivity analysis is an important topic in this book; the concept is revisited in various chapters. In a general sense, a sensitivity analysis aims to evaluate the impact of varying some "input parameters" on some "output response" (Figure 4.9). In decision analysis,

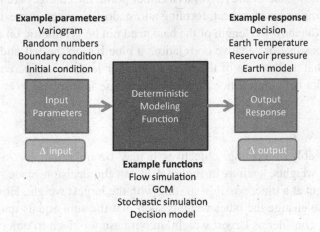

Figure 4.9 General depiction of a sensitivity analysis (GCM = Global Circulation Model).

the output response could be the decision made or the value of a payoff. The relationship between input and output is modeled through a deterministic function, that is, once the input is known, the output is also uniquely known. Note that even a "stochastic simulator" such as Monte Carlo simulation falls in this category, in that case the deterministic function is the probability distribution, the inputs are the parameters of the probability distribution and the random number generated for a given random seed.

In modeling uncertainty as well as in decision analysis, the exact numbers obtained through the analysis and modeling, such as payoffs, probabilities or scores, are often less important than the impact a change in these values have on either the decision made or the model of uncertainty. Indeed, why care about being obsessed with determining a payoff exactly when even a large change of such payoff will not affect the ultimate decision? A sensitivity analysis, even a simple one, allows what matters and what does not to be figured out and can lead to a more focused modeling of the uncertainty about the Earth. This is important, since any modeling of uncertainty requires a context, as such, the actual modeling becomes much simpler when targeted than just modeling uncertainty for the sake of uncertainty.

4.2.4.4.1 Tornado Charts

Tornado charts are used for assessing the sensitivity of a single output variable to each input variable. This entails varying one input variable while leaving all other input variables constant. Varying one input variable one at a time is called "one-way sensitivity." We will cover "multi-way sensitivity" where multiple input variables are varied simultaneously in later chapters.

Tornado charts are visual tools for ranking input parameters based on their sensitivity to a certain response or decision. The input variables are changed one at a time (the others remain fixed) by a given amount on the plus and minus side and the change in response, such as a payoff, are recorded. Often a change of $+/-$ 10% is used; alternatively, the change is made in terms of quantiles, such as given by the interquartile range of the variable. Next, the input variables are ranked in order of decreasing impact on response, in terms of absolute value difference in response for the $+/-$. Using the initial (prior to sensitivity analysis) value of the payoff as a center point, the changes are plotted on a bar chart in descending order of impact, forming a tornado-like shape. An example is shown in Figure 4.10. Note how the length of the bars need not be symmetric on either side. The color indicates positive or negative correlation; a blue bar on the left and red bar on the right indicates that an increase of the input parameter leads to an increase in response. The opposite color bar combination indicates a decrease in response to an increase in the input parameter.

4.2.4.4.2 Sensitivity Analysis in the Presence of Multiple Objectives

In the case of multiple, possibly competing, objectives it may be important to assess the impact of changing the weights, such as in Figure 4.11, on the decision made. Again one can change one weight at a time, possibly starting with the largest weight. However, in doing so, one must also change the other weights such that the sum equals unity. This can be done as follows, consider as largest weight, in our case w_5 given to objective "minimize

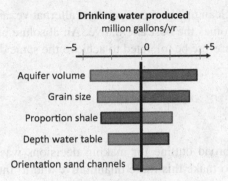

Figure 4.10 Example of a tornado chart.

economic interruption," if this is changed to w_5^{new} then the other weights can be changed as follows:

$$w_1^{new} = (1 - w_5^{new}) \times \frac{w_1}{w_1 + w_2 + w_3 + w_4}, \quad w_2^{new} = (1 - w_5^{new}) \times \frac{w_2}{w_1 + w_2 + w_3 + w_4}, \quad \text{etc.}$$

Each weight is pro-rated according to its contribution to the remaining weights. The impact of changing w_5 on the score values given to each alternative is shown in Figure 4.11. The base case is $w_5 = 0.33$, a change in best score is seen when w_5 drops

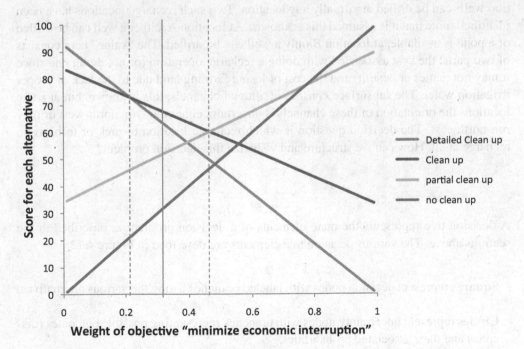

Figure 4.11 Sensitivity of scores on changing the weight on "economic interruption".

to 0.21 when "detailed clean up" becomes the best alternative, and when w_5 increases to 0.46 when no clean becomes the best alternative. An absolute change of 0.13 in weight w_5 on either side can therefore be tolerated to achieve the same decision.

4.3 Tools for Structuring Decision Problems

4.3.1 Decision Trees

In previous sections a broad outline for making decisions was summarized. Here the decision tree as a tool to make this more quantitative and to introduce the language of probability into the actual numerical calculations is discussed. The tree is a visual means for understanding the decision problem as well as organizing the calculations for making an optimal decision.

A fictional but realistic example is considered throughout to illustrate the building of the decision tree. Consider the following scenario (although simplified from the real situation for educational purposes).

Farming near the California coast has led to depletion of the groundwater table and the intrusion of saltwater into the aquifer system, jeopardizing farming activities. A solution is to inject water (recharge) into the subsurface at targeted locations to keep groundwater levels high and hence saltwater intrusion at a minimum. There are two ways of injecting water: either via ponds that filtrate water slowly into the subsurface or through wells stimulating directly the aquifer. Ponds are less expensive but require space, while injection wells can be drilled at virtually any location. Two such recharge locations have been identified (note that it is assumed this is known). At location A, either a well can be drilled or a pond is available, at location B only a well can be drilled. The "value" here consists of two parts: the cost associated with doing a recharge operation (or not doing one since it may not matter or benefit) and the cost of losing farming land due to the lack of proper irrigation water. The subsurface consists of alluvial channels, this is known, but at some locations the orientation of these channels is uncertain, either they run north-west or they run north-east. The decision question is which recharge location to pick or to have no recharge at all. How can we structure and visualize this decision problem?

4.3.2 Building Decision Trees

A decision tree represents the main elements of a decision problem as described in the sections above. The various decision tree elements are described in Figure 4.12:

- **Squares** represent decision nodes with labeled branches listing the various alternatives.

- **Circles** represent uncertainty nodes with branches representing possible outcomes (discrete) and their associated probabilities.

- **Triangles** represent payoffs with values written at the end.

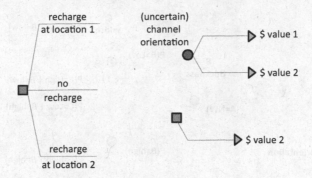

Figure 4.12   Elements of a decision tree.

Using these elements, trees are built from left to right, ordering the sequence of decisions needed in function of time. An example decision tree for our recharge problem is given in Figure 4.13. The first node represents the ultimate decision, that is, whether or not to recharge and if so where to recharge. The subsequent decision is about whether to choose a pond or a well if that choice is available. The uncertainty nodes are listed after the decision node and may or may not be the same for each decision alternative. For example, in location two there is no uncertainty about channel orientation as reflected in the tree of Figure 4.13. We could also have included an additional decision as to whether or not to recharge and then make a decision about location.

Much of this book will be devoted to assigning probabilities to uncertain events as well as ways for calculating values of the end nodes. Indeed, in the example of Figure 4.13 a rather simplified situation is presented, since uncertainty about the Earth subsurface is likely to be more complex than simply that of not knowing exactly channel orientation. In this book, actual 3D Earth models will be built, and due to uncertainty one can build many alternative models constrained to whatever data is available. These models allow calculating responses due to engineering actions or decisions taken in reality such as the

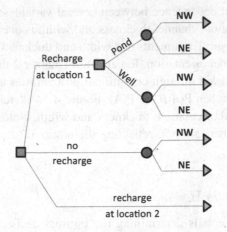

Figure 4.13   Example of a decision tree.

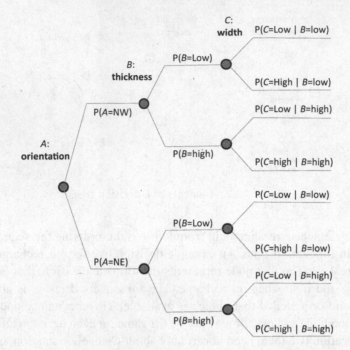

Figure 4.14 Example of hierarchies of uncertain variables in a tree.

decision to recharge or not. Flow simulators can be used to simulate the flow in the sub-surface, thereby simulating the response of a recharge operation. Using economic model calculations, the output of these simulators can then be used to assign a value to each node, for example in this case based on the salt concentration of pumping groundwater out of the aquifer and how that will affect the growth of crops. At this point we will not deal with constructing decision trees when multiple alternative models are the specific representation of uncertainty, for this refer to the chapter on Value of Information.

In simpler cases where uncertainty is simply about a few variables, the decision tree should reflect the mutual dependence between several variables. For example, if, in ad-dition to channel orientation, channel thickness and width is uncertain, then it can be as-sumed from sedimentological arguments that width and thickness are dependent variables but independent of channel orientation. Recall from Chapter 2 that dependency between random variables is modeled through conditional probabilities and that if two variables A and B are independent then $P(A|B) = P(A)$. Figure 4.14 sketches a plausible situation for the case with channel orientation, thickness and width. Note how some probabilities are conditional while others are not, reflecting the nature of dependency between these various variables.

4.3.3 Solving Decision Trees

Solving a decision tree entails determining the optimal decision for the leftmost (i.e., ultimate) decision by maximizing an expected value. To achieve this, the tree is solved from right to left using the following method:

1 Select a rightmost node that has no successors.

2 Determine the expected payoff associated with the node:

   a If it is a decision node: select the decision with highest expected value

   b If it is a chance node: calculate its expected value.

3 Replace the node with its expected value.

4 Go back to step 1 and continue until you arrive at the first decision node.

Consider the drinking water contamination example introduced at the beginning of the chapter. Figure 4.15 shows the decision tree which can be read from left to right as follows: the decision is whether to clean up or not, in case of a clean up decision, the government needs to pay the cost, which is $15 million (hence a value of −15). In case of a

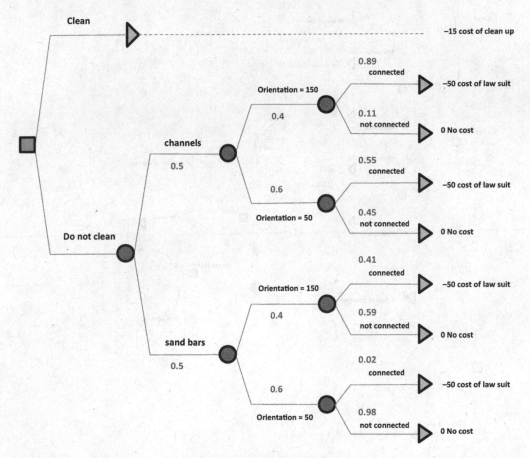

Figure 4.15   Decision tree for the groundwater contamination problem (red are probabilities for that node).

decision to not clean up the site, there is a potential for a law suit. This is subject to chance because the law suit will only happen when the drinking water is polluted, which in turn will happen when the geology of the subsurface is unfavorable. Such an "unfavorable" situation occurs when a subsurface "connection" exists between the contaminant source and the well. A connection means that there is a path through which contaminants can travel from one point to another. In subsequent chapters, how to calculate the probability of such connection occurring given the geological uncertainty is discussed. For now, assume two scenarios, "connected" and "not-connected," and assume the probability for all possible cases of geological uncertainty is known. Such uncertainty presents itself at two levels (and they are assumed to be hierarchical, so the order in the tree matters): uncertainty in geological scenario (presence of channels vs presence of bars) and uncertainty in orientation for a given geological scenario, both being binary and discrete. The cost of the law suit is set at $50 million (although this is usually uncertain as well).

Starting now from right to left, the tree is "solved" by calculating the expected values for each branch (Figure 4.16) until the decision node is reached. The optimal decision is then to clean up, since this has the smallest cost (largest value).

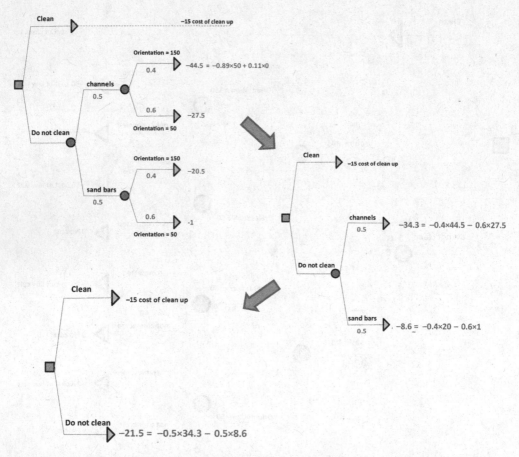

Figure 4.16 Solving the decision tree from right to left for the groundwater contamination problem.

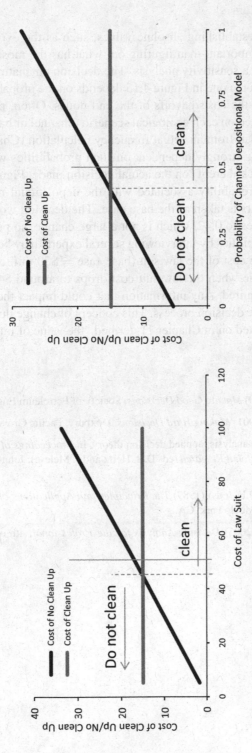

Figure 4.17 Sensitivity of the decision made for (left) changing the cost of law suit and (right) the probability associated with the channel depositional model.

4.3.4 Sensitivity Analysis

As discussed previously, establishing absolute values, such as the expected values calculated in a tree, is less important than figuring out what has the most impact on these values. This is the goal of a sensitivity analysis. The decision emanating from solving a given decision tree such as the one in Figure 4.15 depends on the probabilities associated with uncertain events and the costs/payoffs of the end-nodes. Often, prior probabilities such as the probability stated on each geological scenario (channel or bars) emanate from expert judgments rather than from an actual frequency calculation (Chapter 2). Figuring out whether the decision is strongly dependent on such probabilities will provide some confidence (or possibly lack thereof) on the actual decision made. Figure 4.17 shows an example of varying the probability associated with the depositional model, if a 50/50 chance for channel and bars is taken as the base case. The decision would change if the channel probability changes to 0.42, which is not a large change, so putting some extra effort on these prior probabilities by interviewing several experts may be worth the effort.

In another example, the cost of the law suit (base case = $50 million) can be varied. The decision would change when the law suit cost drops to around $46 million, hence only a small change is required. Any information that could impact these numbers may therefore be valuable to the decision process. This concept of change due to collecting or using more data is elaborated on in Chapter 11, termed "the value of information."

Further Reading

Bratvold, R. and Begg S. (2010) *Making Good Decisions*, Society of Petroleum Engineers, Austin, TX.

Clemen, R.T. and Reilly, T. (2001) *Making Hard Decisions*, Duxbury, Pacific Grove, CA.

Howard, R.A. (1966) Decision analysis: applied decision theory, in *Proceedings of the Fourth International Conference on Operational Research* (eds D.B. Hertz and J. Melese), John Wiley & Sons, Inc., New York, NY, pp. 55–71.

Howard, R.A. and Matheson, J.E. (eds) (1989) *The Principles and Applications of Decision Analysis*, Strategic Decisions Group, Menlo Park, CA.

McNamee, P. and Celona, J. (2005) *Decision Analysis for the Professional*, 4th edn, SmartOrg, Inc., Menlo Park, CA.

5

Modeling Spatial Continuity

Modeling spatial continuity is critical to the question of addressing uncertainty, since a spatial model of the properties being studied will lead to a different assessment of uncertainty compared to assuming everything is random.

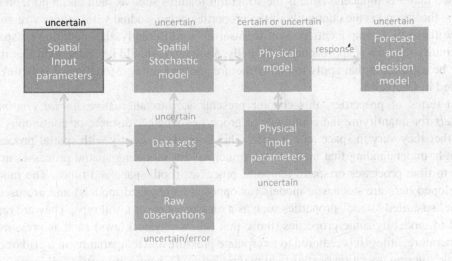

5.1 Introduction

Variability in the Earth Sciences manifests itself at many levels: properties vary in space and/or time with alternating high and low values. The spatial distribution, that is, the characteristics of how these highs and lows vary is important for many engineering and decision making problems. Mathematical models are needed to characterize this spatial distribution then create Earth models (i.e., assign properties to a grid) that reflect the conceptualized spatial distribution. However, due to incomplete information this conceptual model is subject to uncertainty. Moreover, a mathematical model or concept can only partially capture true variability.

Modeling Uncertainty in the Earth Sciences, First Edition. Jef Caers.

The non-randomness of Earth Sciences phenomena entails that values measured close to each other are more "alike" than values measured further apart, in other words a spatial relationship exists between such values. The term "spatial relationship" incorporates all sorts of relationships, such as relationships among the available spatial data or between the unknown values and the measured data. The data may be of any type, possibly different from that of the variable or property being modeled. Therefore, in order to quantify uncertainty about an unsampled value, it is important to first and foremost quantify that spatial relationship, that is, quantify through a mathematical model the underlying spatial continuity. In this book such a model is termed a "spatial continuity model". The simplest possible quantification consists of evaluating the correlation coefficient between any datum value measured at locations $\mathbf{u} = (x,y,z)$ and any other measured a distance h away. Providing this correlation for various distances h will lead to the definition of a correlogram or variogram, which is one particular spatial continuity model discussed.

In the particular case of modeling the subsurface, spatial continuity is often determined by two major components: one is the structural features such as fault and a horizon surfaces; the second is the continuity of the properties being studied within these structures. Since these two components present themselves so differently, the modeling of spatial continuity for each is approached differently, although it should be understood that there may be techniques that apply to both structure and properties. Modeling of structures is treated in Chapter 8.

In terms of properties, this chapter presents various alternative *spatial continuity models* for quantifying the continuity of properties whether discrete or continuous and whether they vary in space and or time. This book deals mostly with spatial processes though, understanding that many of the principles for modeling spatial processes apply also to time processes or spatio-temporal processes (both space and time). The models developed here are stochastic models (as opposed to physical models) and are used to model so-called "static" properties such as a rock property or a soil type. They are rarely used to model dynamic properties (those that follow physical laws) such as pressure or temperature, unless it is required to interpolate pressure and temperature on a grid or do a simple filtering operation or statistical manipulation. Dynamic properties follow physical laws and their role in modeling uncertainty is the topic of Chapter 10.

Three spatial models are presented in this chapter: (i) the correlogram/variogram model, (ii) the object (Boolean) model, and (iii) the 3D training image models. As with any mathematical model, "parameters" are required for such models to be completely specified. The variogram is a model that is built based on mathematical considerations rather than physical ones. While the variogram may be the simplest model of the set, requiring only a few parameters, it may not be easy to interpret from data when there are few, nor can it deliver the complexity of real spatially varying phenomena. Both the object-based and the training image-based models attempt to provide models from a more realistic perspective, but call for a prior thorough understanding and interpretation of the spatial phenomenon and require many more parameters. Such interpretation is subject to a great deal of uncertainty (if not one of the most important one).

5.2 The Variogram

5.2.1 Autocorrelation in 1D

The observation that is made at a certain time instant t_i is generally not independent from the time observation at, for example, the next time instant t_{i+1}. Or, in space, the observation made at a position x_i is not independent of the observation made at a slightly different position x_{i+1}. For prediction purposes, it is often very important to know how far the correlation or association extends or what the exact nature of this association is. We would like to design a measure of association for each time or space interval. We expect that, for small time intervals, the two events separated by this interval are very well associated. However, as the time interval increases, we expect the association to decrease.

In Chapter 2 a measure of linear association between two variables X and Y, namely the correlation coefficient, was defined. Here, this idea is extended to one variable measured at different instances in time. Recall the equation for the correlation between X and Y:

$$r = \frac{1}{n-1} \sum_{i=1}^{n} \left(\frac{x_i - \overline{x}}{s_x} \right) \cdot \left(\frac{y_i - \overline{y}}{s_y} \right)$$

The correlation measured between two variables X and Y is applied to the correlation measured between the same variable Y but measured at different time instances. We will consider time instances that are Δt apart.

Using a sliding scheme (Figure 5.1), the pair $[Y(t), Y(t + \Delta t)]$ are moved over the time axis, each time t, recording the values of $y(t)$ and $y(t + \Delta t)$. The pair $[y(t), y(t + \Delta t)]$ represents one single point in the scatter plot of Figure 5.1. $n(\Delta t)$ is the number of pairs of data found by applying this slide rule for a given Δt. It is observed that as Δt increases, fewer data pairs are found simply because the time series has a limited record. The complete scatter plot in Figure 5.1 is then used to calculate the correlation coefficient

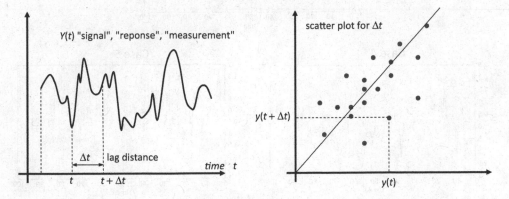

Figure 5.1   A 1D time series (left). Calculating the correlation coefficient for a given lag distance or time interval Δt (right).

for that value of Δt. Calculating the correlation coefficient for various intervals Δt and plotting $r(\Delta t)$ versus Δt results in the correlogram or autocorrelation plot.

What does the autocorrelation plot measure?

- When $\Delta t = 0$: the value $r = 1$ is retained. Indeed, $y(t)$ is perfectly correlated with itself.

- When $\Delta t > 0$: One expects the correlation r to become smaller. Indeed, events that are farther spread apart in time or space are less correlated with each other.

A few simple example cases are shown in Figure 5.2a and 5.2b. In *Case 1*, it can be observed how the correlation coefficient becomes approximately zero at twenty time

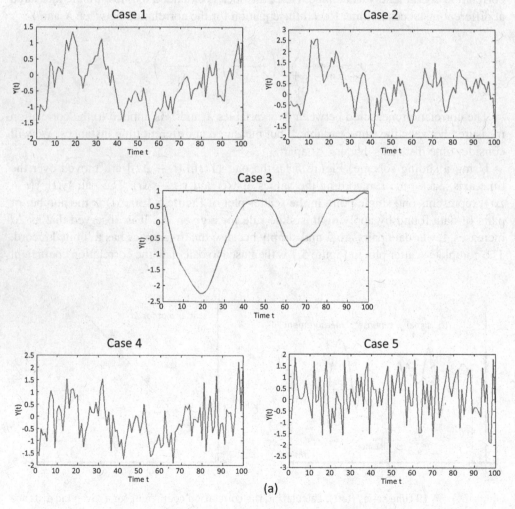

Figure 5.2 (a) Example time series; (b) correlograms corresponding to the time series.

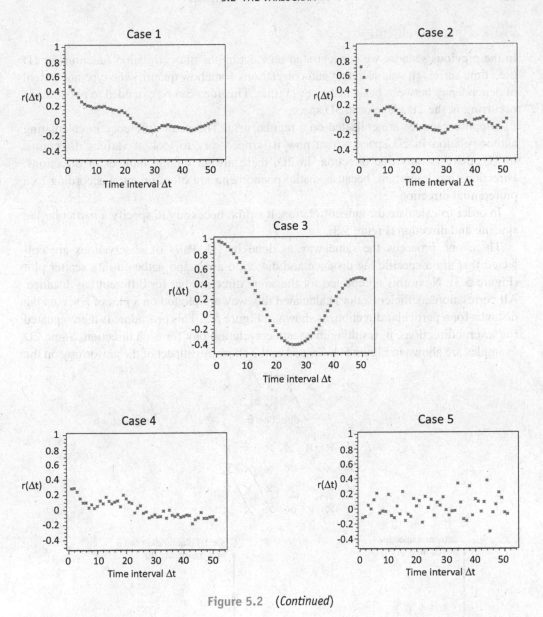

Figure 5.2 (*Continued*)

units. This length is termed the correlation length. In *Case 2* the correlation length is approximately 10 units. In *Case 3* periodicity of the time series, which is also present in the correlogram, is observed. In *Case 4* discontinuity for small distance Δt (the correlation drops from unity rapidly to 0.4) is observed; this usually means that there is either a measurement error or a small variation that the sampling did not detect. *Case 5* shows perfectly uncorrelated time events, hence the correlogram fluctuates around zero for all intervals.

5.2.2 Autocorrelation in 2D and 3D

In the previous section, we concentrated on calculating autocorrelation functions in 1D (i.e., time series). It was seen that autocorrelations somehow quantify the type and extent of dependency between basic instances in time. This idea can be extended to phenomena occurring in the 2D plane or in 3D space.

Suppose samples are collected on a regular grid. The main difference in calculating autocorrelation in 2D space is that now it is necessary to look at various directions. In 1D, there is only one direction. In 2D, there are an infinite number of directions. Directions are important, because spatial phenomena are often oriented according to a preferential direction.

In order to calculate the autocorrelation, it is first necessary to specify a particular lag spacing and direction (Figure 5.3).

Then, one proceeds the same way as done in 1D. Pairs of observations are collected that are a specific lag distance and direction away and gathered in a scatter plot (Figure 5.3). Next, this is repeated for the *same* direction but for different lag distance. All correlation coefficient values calculated this way are plotted on a plot of r versus lag distance for a particular direction, as shown in Figure 5.3. This procedure is then repeated for several directions, θ, resulting in an autocorrelation plot for each direction. Some 2D examples are shown in Figure 5.4. Cases 1 and 2 show the impact of the anisotropy on the

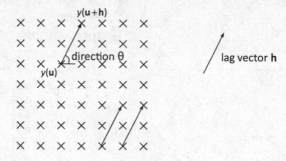

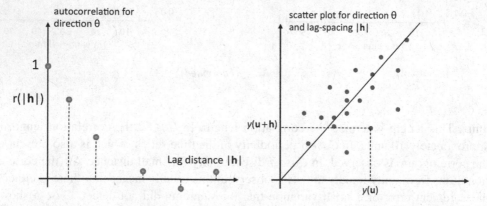

Figure 5.3 Direction and lag-spacing between two samples on a regular grid, corresponding scatter plot and construction of the correlogram or autocorrelation plot for that direction. **u** is a coordinate in space.

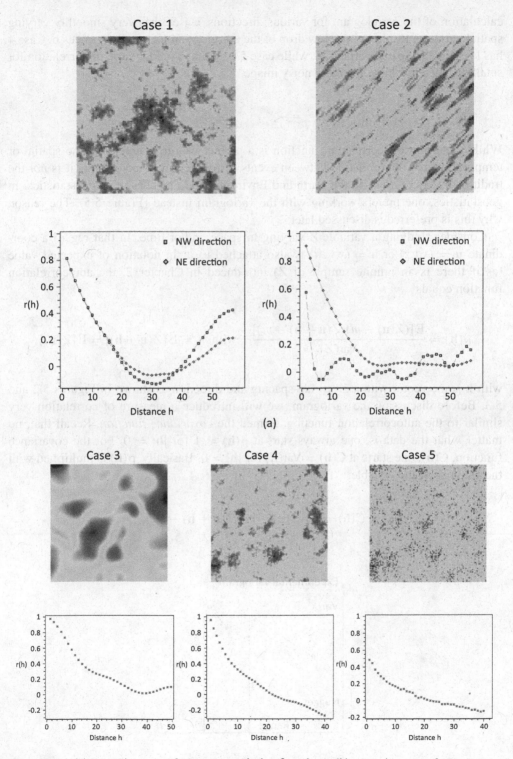

Figure 5.4 (a) Example cases of 2D autocorrelation functions; (b) example cases of 2D autocorrelation functions.

calculation of the correlogram for various directions. Case 3 has very smoothly varying spatial variation resulting in a slow drop of the correlogram for small distances h. Case 4 has less smooth spatial variability, while case 5 contains a sudden drop of correlation for small distance resulting in a more noisy image.

5.2.3 The Variogram and Covariance Function

While the empirical correlation function is a perfect way to characterize the spatial or temporal degree of correlation between events distributed in space or time, it is *not* the traditional way to do so in the Earth and Environmental Sciences or in geostatistics. In geostatistics, one prefers working with the variogram instead (Figure 5.5). The reason why this is preferred is discussed later.

Consider studying a variable Z varying in space and/or time. In that regard a coordinate $\mathbf{u} = (x,y,z)$ or $\mathbf{u} = (x,y,z,t)$ is also attached to it. In notation of expected value (as if there is an infinite sample of Z) introduced in Chapter 2, the autocorrelation function equals

$$\rho(\mathbf{h}) = \frac{E\left[(Z(\mathbf{u}) - m)(Z(\mathbf{u} + \mathbf{h}) - m)\right]}{\text{Var}(Z)} \quad \text{with } m = E\left[Z(\mathbf{u} + \mathbf{h})\right] = E\left[Z(\mathbf{u})\right]$$

with $\mathbf{h}$ some vector with a given lag-spacing and direction as shown in Figures 5.2 and 5.3. Before discussing the variogram, we will introduce a measure of correlation very similar to the autocorrelation function, termed the *covariance function*. Recall that, no matter what the data is, one always start at $\rho(\mathbf{h}) = 1$ for $|\mathbf{h}| = 0$. For the covariance function, $C(\mathbf{h})$, one starts at $C(\mathbf{h}) = \text{Var}(Z)$ for $|\mathbf{h}| = 0$. Basically, $\rho(\mathbf{h})$ is multiplied with the variance of the variable:

$$C(\mathbf{h}) = E\left[(Z(\mathbf{u}) - m)(Z(\mathbf{u} + \mathbf{h}) - m)\right]$$

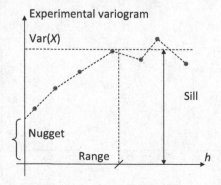

Figure 5.5 Elements of an experimental variogram.

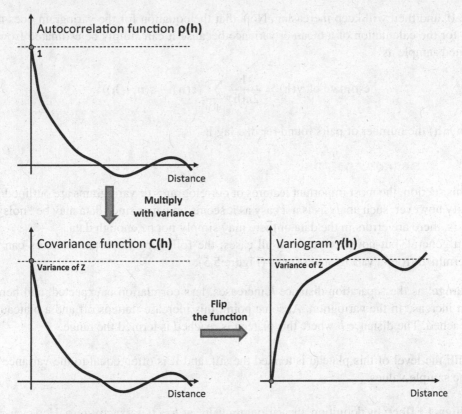

Figure 5.6 Relationship between autocorrelation, covariance and variogram functions.

While the autocorrelation function does not contain any information on the variance of the phenomenon under investigation, the covariance function does. Finally, the (semi)-variogram[1] is defined as

$$\gamma\,(\mathbf{h}) = \mathrm{Var}(Z) - C\,(\mathbf{h})$$

which is equivalent to

$$\gamma\,(\mathbf{h}) = \frac{1}{2}\mathrm{E}\left[(Z(\mathbf{u}) - Z(\mathbf{u} + \mathbf{h}))^2\right]$$

One might wonder what this is all good for. After all, we first did a simple multiplication and then flipped the function (Figure 5.6). Suppose, for example, that the variance is really large; in fact it gets larger as more data are gathered over an increasingly larger area (it may be an indication that the variance is not very well defined). In that case, the covariance function, which starts at the variance, would become unstable, it continuously changes. The variogram does not have this problem. It would simply start at zero for

---

[1] Variogram and (semi)-variogram are often used intertwined, although the latter is half the former.

$\mathbf{h} = 0$, and then will keep increasing. Note that the equation for the variogram does not call for the calculation of a mean or variance because it can simply be estimated from a limited sample as

$$\text{estimate of } \gamma(\mathbf{h}) = \frac{1}{2n(\mathbf{h})} \sum_{\text{all } \mathbf{u}} (z(\mathbf{u}) - z(\mathbf{u} + \mathbf{h}))^2$$

with $n(\mathbf{h})$ the number of pairs found for that lag $\mathbf{h}$.

5.2.4 Variogram Analysis

In this section, the most important features of correlograms or variograms are outlined. In reality however, such analysis is not easy as it seems on paper since data may be "noisy", that is, there are errors in the data or there may simply not be enough data.

In general, but not necessarily in all cases, the following important features can be determined from a variogram estimate (Figure 5.5):

- **Range**: as the separation distance h increases, less correlation is expected, and hence an increase in the variogram. At some point, this increase flattens off and a plateau is reached. The distance h where this plateau is reached is termed the range.

- **Sill**: the level of this plateau is termed the sill, and it is often equal to the variance of the sample values.

- **Nugget Effect**: by definition, the variogram value at $h = 0$ is exactly zero. However, for small h, a sudden jump in the variogram value is often observed. This jump is termed the nugget effect.

This nugget effect is often due to small scale variability that is not sampled because the distance between the samples is too large. Historically, the term refers to the small scale variability that is caused in gold mines when a sample contains a gold nugget (a very short scale variation of gold grade). Another reason for a nugget effect is error (noise) in the measurements. If these errors act like random noise, then a nugget effect will be observed.

5.2.4.1 Anisotropy

We showed earlier that the range (correlation length) can vary according to the direction in which the variogram is calculated. This makes sense since the Earth may vary differently in different directions. For example, layers in the Earth are more continuous in the horizontal direction than in the vertical direction.

5.2.4.2 What is the Practical Meaning of a Variogram?

A variogram measures the *geological distance* between any two points in space. Recall that a Euclidean distance simply measures the distance between any two locations in a Cartesian space.

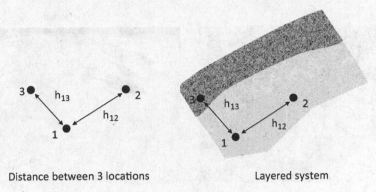

Figure 5.7 A variogram distance versus a Euclidean distance.

Figure 5.7 explains the difference between Euclidean and "geological" distance: suppose the geology of a region is layered, then it appears that there is clear anisotropy in the 45° diagonal direction. Hence, samples 1 and 2 can, at least geologically, be considered as more similar (hence closer) than samples 1 and 3. This degree of similarity is measured by the variogram. It would be expected that:

$$\gamma\,(\mathbf{h}_{13}) > \gamma\,(\mathbf{h}_{12}) \text{ even though } \mathbf{h}_{12} > \mathbf{h}_{13}.$$

5.2.5 A Word on Variogram Modeling

Simply calculating the variogram in 1, 2 or 3D is not enough, it is necessary to provide a "model", just like a set of numbers is not a model pdf, it is only an empirical pdf; it is necessary to complete the pdf by interpolation and extrapolation rules as outlined in Chapter 2. Variogram modeling is not trivial since there are some more restrictions on what can be done, but the idea is the same: interpolation of the calculated variogram with a smooth curve. This topic is for more advanced geostatistics books and this work is typically carried out by geostatisticians. Throughout the further chapters it is assumed that such a model is available, that is, that we can evaluate the variogram for any lag distance and direction.

5.3 The Boolean or Object Model

5.3.1 Motivation

Variograms are crude descriptions of actual spatial phenomena. Since the variogram captures spatial continuity by considering sample values taken two at a time, it cannot capture well complex spatial phenomena, such as shown in Figure 5.8 and 5.9. Indeed, a nugget value, a range (or set of ranges per direction) and a sill could not describe the complex sinuous variation of a channel or the growth of carbonate mounds and reefs that may need hundreds of parameters for a complete description.

Since all cases in Figure 5.9 are actual naturally occurring systems, subject to physical laws, one idea is to simply "simulate" the genesis (deposition, growth, evolution) of these

Real   Simulated

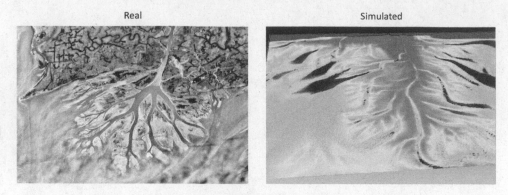

Figure 5.8   A real phenomena versus a process simulation of it.

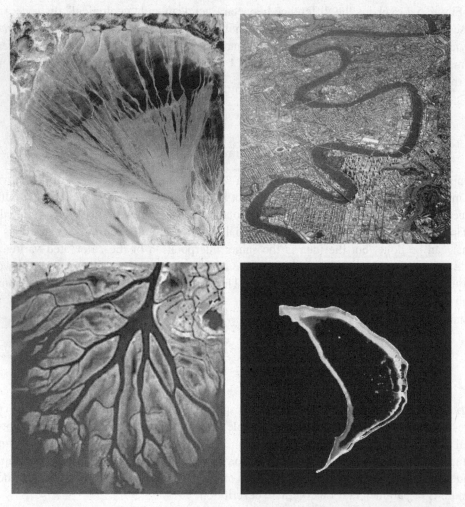

Figure 5.9   An alluvial fan (top left), a meandering river (top right), a delta (bottom left), an atol (bottom right).

systems on a computer. For example, it is possible to simulate how a delta dumps sand and silt material over the delta plain by looking at rules of deposition and erosion coupled with physical laws of turbulent fluid flow. Figure 5.8 gives an example. In geological modeling these models are often termed "process" models.

However, such process simulation may need days, sometimes weeks of computation time. Referring to the discussion in Chapter 3 on deterministic models, such models may be physically realistic but may not have much quantitative prediction power for two reasons: (1) they do not reflect uncertainty about the phenomena, (2) they may be hard to calibrate to data, particularly in subsurface modeling where data from wells and geophysical surveys are available. However, the result of a process model may provide a wealth of information about the style of spatial variation, the architecture of sediments and so on. Such information can be used by simpler, nonphysical models to mimic, not through actual simulation of processes but through stochastic simulation, the style of spatial continuity present. One such model is the Boolean model. If spatial continuity presents itself through objects, then a Boolean model describes the geometry, dimension and interaction of these objects. A Boolean Earth model is then an actual representation of this model description in 3D. Figure 5.10 shows that a Boolean Earth model can mimic fairly well the style of spatial continuity observed in a process model.

5.3.2 Object Models

Object models (also termed "Boolean models") were introduced to overcome some of the limitations of the variogram-based tools by importing realistic shapes and associations into a model. Crisp curvilinear shapes are often hard to model with cell-based techniques. The object simulation approach consists of dropping directly onto the grid a set of objects

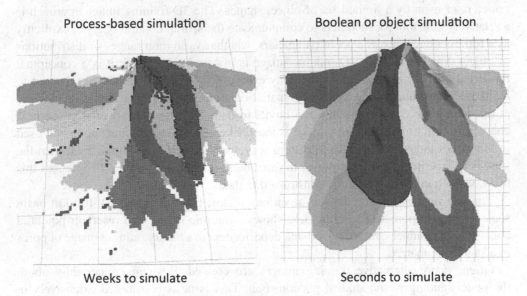

Process-based simulation   Boolean or object simulation

Weeks to simulate   Seconds to simulate

Figure 5.10 A process model versus a simulation of a Boolean model.

representing different categories (rock types, sedimentary facies, and fractures). These objects are then moved around to match the data by Markov chain simulation, which is discussed in Chapter 6. This technique has mostly been used to model sedimentary objects in reservoirs or aquifers, but many other applications can be envisioned, such as for example the simulation of gold veins.

Before "simulating" the objects, it is necessary to define the object model. The first task is to establish the various types of objects (sinuous, elliptic, cubic) and their dimensions (width, thickness or width to thickness ratio, vertical cross section parameters, sinuosity, etc.) which can be constant or varying according to a user specified distribution function. Next it is necessary to specify their mutual spatial relationship: erosion of one object by another, embedding, and attraction/repulsion of objects.

For example, in the case of channel type system (fluvial or submarine) various sources of information can be considered to define the object model:

- Outcrop studies of analog systems are probably the best source of information, although there is a possibility of bias that comes up when inferring 3D object properties from 2D outcrops (smaller 3D objects are less likely to occur in 2D sections).

- Data from the site itself may provide information on object geometry, or may at least help in relating object shape parameters from outcrop data to the objects shapes being simulated.

5.4 3D Training Image Models

In many cases, it is not possible to capture spatial complexity by a few variogram parameters or even by a limited set of object shapes. The 3D training image approach is a relatively new tool for modelers to communicate the spatial continuity style, explicitly as a full 3D image, not as a set of parameters, whether variogram ranges or distributions of object dimensions. The 3D training image is not an Earth model; it is a conceptual rendering of the major variations that may exist in the studied area. The aim is then to build 3D Earth models that mimic the spatial continuity of the 3D training image, and at the same time constrain such Earth model to data. This topic is covered in the next chapter. In this way the training image is much like a "texture mapping" approach used in the games industry. A particular pattern is presented, such as in Figure 5.11, then the trick is to randomize this pattern over the area being modeled. In the Earth Sciences this must be done in 3D as well as be constrained to data.

Training images may be defined at various scales, from the large 10–100 km basin scale to the μm pore scale. Figure 5.12 shows a training image of a reservoir potential consisting of channel sand with overbank deposits next to a binary training image of pores in a sandstone matrix.

Often, many alternative training images are created, reflecting uncertainty about the understanding of the studied phenomenon. This issue is considered extensively in Chapters 9 and 10.

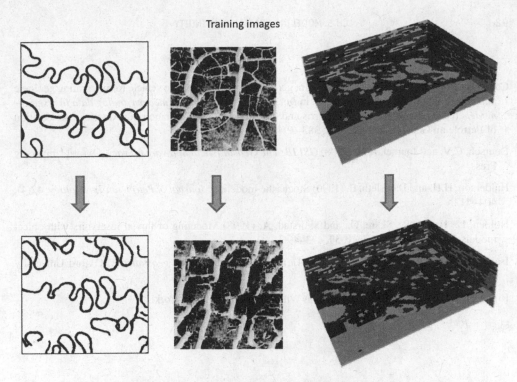

Figure 5.11 A few example training images and Earth models produced from them exhibiting similar patterns.

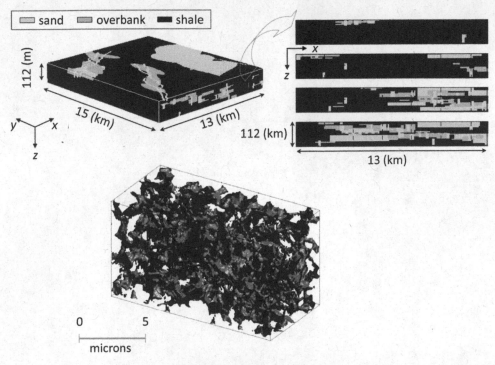

Figure 5.12 3D Training image example at the basin scale and pore scale (red = pore space).

Further Reading

Caers, J. and Zhang, T. (2004) Multiple-point geostatistics: a quantitative vehicle for integrating geologic analogs into multiple reservoir models, in *Integration of outcrop and modern analog data in reservoir models* (eds G.M. Grammer, P.M. Harris and G.P. Eberli), AAPG Memoir 80, American Association of Petroleum Geologists, Tulsa, OK, 383–394.

Deutsch, C.V. and Journel, A.G. (1998) *GSLIB: The Geostatistical Software Library*, Oxford University Press.

Halderson, H.H. and Damsleth, E. (1990) Stochastic modeling. *Journal of Petroleum Technology*, **42**(4), 404–412.

Holden, L., Hauge, R., Skare, Ø., and Skorstad, A. (1998) Modeling of fluvial reservoirs with object models. *Mathematical Geology*, **30**, 473–496.

Isaaks, E.H. and Srivastava, R.M. (1989) *An introduction to applied geostatistics*, Oxford University Press.

Ripley, B.D. (2004) *Spatial statistics*, John Wiley & Sons, Inc., New York.

6
Modeling Spatial Uncertainty

The set of models that can be generated for a given (fixed) spatial continuity model represents a type of uncertainty that is termed spatial uncertainty or model of spatial uncertainty. It is important to recognize that this is only one component in the large model of uncertainty that is being built in this book.

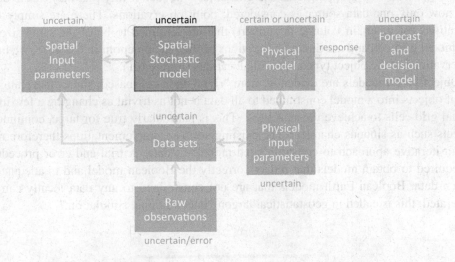

6.1 Introduction

A variogram model, Boolean model or 3D training image provides a model for the spatial continuity of the phenomenon being studied. This was the topic of the previous chapter. In this chapter methods for generating an Earth model that reflects what is captured in this spatial continuity model are discussed. Recall that an Earth model is simply a 3D grid filled with one or more properties. It will be seen that this creation is not a unique process; many Earth models that are reflective of the same spatial continuity can be created. The set of models that can be generated for a given (fixed) spatial continuity model

Modeling Uncertainty in the Earth Sciences, First Edition. Jef Caers.
© 2011 John Wiley & Sons, Ltd. Published 2011 by John Wiley & Sons, Ltd.

represents a type of uncertainty that is termed spatial uncertainty or model of spatial uncertainty. It is important to recognize that this is only one component in the large model of uncertainty that is being built in this book. If conceptually the Earth is understood as infinitely long one-dimensional layers, then only one model can be created: a layered model. If channel-type structures are assumed to exist and modeled in a Boolean model, then there are many ways to spatially distribute these channels in 3D space in a way that is geologically realistic following the geological rules specified by the Boolean model. In this chapter various techniques to accomplish this are discussed, then, in the next chapter, it is shown how this spatial uncertainty can be further constrained with various sources of data.

6.2 Object-Based Simulation

As discussed previously, object models allow the shape of objects occurring in nature to be realistically represented. The aim of an object-based algorithm is to generate multiple Earth models by dropping onto the grid objects in such a way that they fit the data. For now only one data source is considered: point observations. These are simply observations (of a certain volume or support) that are assumed to be of the same size as the model grid cell and are direct observations of the phenomenon. In this case, we have observations of the object type at specific locations (Figure 6.1).

Object-based models are evidently more "rigid" than cell-based, since "morphing" a set of objects into a model constrained to all data is not as trivial as changing a few individual grid cells to achieve the same task. This is particularly true for large, continuous objects such as sinuous channels. The existing software implementations therefore rely on an iterative approach to constrain such models to data. A trial-and-error procedure is required to obtain models that reflect correctly the Boolean model and to adequately match data. Boolean Earth models that are not constrained to any data locally can be generated; this is called in geostatistical jargon "unconditional simulation."

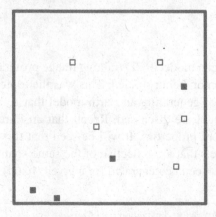

Figure 6.1   Point data with two categories.

For unconditional Boolean simulation with one object type, an Earth model can be simply simulated without iteration by placing the objects randomly in the Earth model grid and continuing until the desired proportion of objects is achieved. However, when multiple interacting objects need to be placed in the same grid or when the placement of objects need to follow certain rules, it is often necessary to "iterate," for example by moving the objects around until the rules expressed in the Boolean model are respected. Iteration is required since it would be too hard to achieve this by a stroke of luck in the first iteration. The same holds when there are data that need to be matched. These iterative approaches start by generating an initial model that follows the pre-defined shape description but does not necessarily fit the local data or follow all the rules. The most critical part in making the iterative approach successful is the way in which a new object model is generated, perturbed from the initial or current one. The type of perturbation performed will determine how efficient the iterative process is, how well the final object model matches the data and how well the pre-defined parameterization of object shapes is maintained. One iteration step in this iterative scheme consists of:

- *proposing* a perturbation of the current 3D Earth model;

- *accepting* this perturbation with a certain probability α: this means that there is some chance, namely $1 - \alpha$, that a model perturbation that improves the data matching will be rejected. This is needed in order to cover as much as possible all possible spatial configurations of objects that match equally well the data.

Several theories on so-called Markov chains (a process whereby the next iteration accounts only for the previous one) have been developed on defining the optimal perturbation and on determining the probability α at each iteration step and are specific to the type of objects present. These are not discussed in this book, as this is more advanced material. Importing objects and arbitrarily morphing them to match the data could be done easily. But then odd or unrealistic shapes may be generated. The key lies in determining values for α that achieve two goals: (1) matching data and (2) reflecting the predefined Boolean model.

Regardless of the numerous smart implementations, the most challenging obstacle in using object-based algorithms lies in the mismatch between the object parameterization and the actual data. Some discrepancy exists between the simplified geometrical shapes and actual occurrence in nature; reducing that discrepancy may call for considerable CPU demand due to long iterations. It is often not possible to predict the level of discrepancy prior to starting the object-based algorithms. While the object-based approach is a general approach in the sense that any type of object could be modeled, the iterative approaches to constraining such models to data are usually object specific.

A few examples of typical rules and geometries used on object simulation are shown in Figure 6.2. Case 1 has a simple elliptical object randomly distributed with a proportion of 30%. Case 2 shows the same elliptical object, but having a proportion that varies in space according to the map shown in Figure 6.2 (red color meaning higher proportion). Case 3 shows an Earth model constrained to the data of Figure 6.1. Case 3 shows

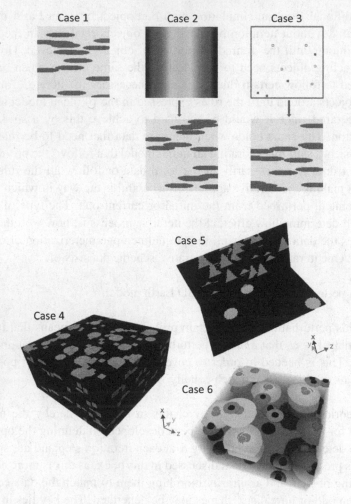

Figure 6.2 A few example cases of object simulations.

randomly placed objects consisting of two parts (also termed elements). Case 4 shows two objects (the red and the green) that never touch each other, while the yellow and green always touch each other. Case 5 shows a complex stacking of an object that consists of two parts (termed elements). Case 6 shows a complex 3D object simulation that combines various features of the previous cases. This Earth model is reflective of the growth of two types of carbonate mounds consisting of an inner core with different rock properties as the outer shell.

6.3 Training Image Methods

6.3.1 Principle of Sequential Simulation

Most of the geostatistical tools for modeling spatial uncertainty are cell based (pixel based). The statistical literature offers many techniques for stochastically simulating 3D

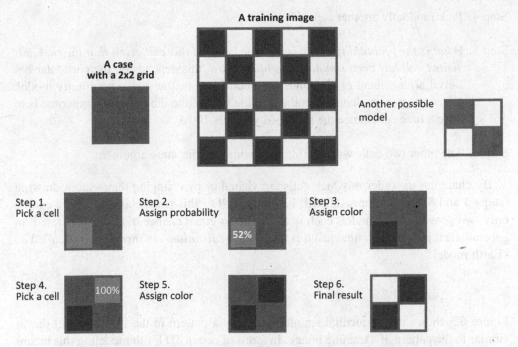

Figure 6.3 Step-by-step description of sequential simulation for a 2 × 2 grid.

Earth models. The size of Earth models (millions of cells) prevents most of them of being practical either because of CPU or RAM (memory) issues. One methodology that does not suffer from these limitations is sequential simulation. The theory behind the sequential simulation approach is left to advanced books in geostatistics, the how-and-why it works can be intuitively explained. In a nutshell: starting from an empty Cartesian grid a model is built one cell at a time by visiting each grid cell along a random path, assigning values to each grid cell, until all cells are visited. Regardless of how grid cell values are assigned, the value assigned to a grid cell depends on the values assigned to all the previously visited cells along the random path. It is this sequential dependence that forces a specific pattern of spatial continuity into the model.

Consider the example in Figure 6.3. The goal is to generate a binary model for a 2 × 2 grid that displays a checkerboard pattern similar to the training image shown at the top of Figure 6.3. The sequential simulation of this 2 × 2 grid proceeds as follows:

Step 1. Pick any of the four cells.

Step 2. *What is the probability of having a blue color in this cell?* Since there are not yet any previously determined cells, that probability is equal to the overall estimated percentage of blue pixels, in this case $13/25 = 52\%$ because there are thirteen blue cells and twelve white cells in the training image.

Step 3. Given the 52% probability, determine whether by random drawing a blue or white color will be given this cell. Suppose the outcome of this random drawing is blue.

Step 4. Pick randomly another cell.

Step 5. *What is the probability of having a blue color in this cell given that the previous visited cell has been simulated as blue?* Now, this depends on the particular believed arrangement of blue and white cells, that is, the spatial continuity model. Given the pattern depicted in the training image, the only possible outcome is to have a blue cell, hence the probability equals 100%.

Step 6. The other two cells will be simulated white for the same argument.

By changing the order in which cells are visited or by changing the random drawing (steps 3 and 5) a different result will be obtained. For this particular case, there can be only two possible final models each with an almost equal chance of being generated. In geostatistical jargon, each final result is termed a *"realization"*. In this book it is called an "Earth model."

6.3.2 Sequential Simulation Based on Training Images

Figure 6.3 shows that sequential simulation forces a pattern in the 2×2 model that is similar to the pattern of a training image. In terms of actual 3D Earth modeling this means that the training image expresses the pattern one desires the Earth model to depict. The sequential procedure explained step-wise above is essentially similar for complex 3D models with large 3D training images. At each grid cell in the Earth model, the probability of having a certain category, given any previously simulated categories, is calculated. Take for example, the situation in Figure 6.4. The probability of the central cell being

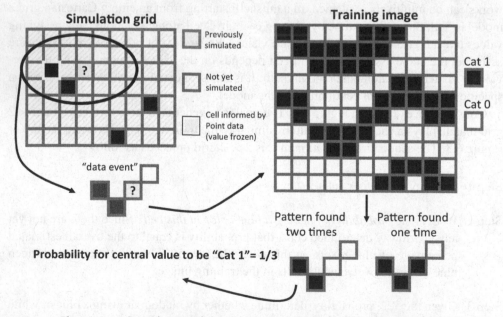

Figure 6.4   Scanning a training image to obtain a conditional probability.

channel sand given its specific set of neighboring sand and no-sand data values, is calculated by scanning the training image in Figure 6.4 for "replicates" of this data event: three such events are found of which one yields a central sand value, hence the probability of having sand is 1/3. By random drawing, a category is assigned. This operation is repeated until the grid is full. This procedure results in a simulated Earth model which will display a pattern of spatial continuity similar to that depicted in the training image.

In actual cases, data provide local constraints on the presence of certain values/categories. In sequential simulation, such constraints are handled easily by assigning (freezing) category values to those grid cells that have point data (= data that inform directly the cell value) (Figure 6.4). The cells containing such constraints are never visited and their values never re-considered. The sequential nature of the algorithm forces all neighboring simulated cell values to be consistent with the data. Unlike object-based algorithms, sequential simulation methods allow constraining to well data in a single pass over all grid cells, no iteration is required.

6.3.3 Example of a 3D Earth Model

Consider the example of modeling rock types in a tidal dominated system further illustrates the concept of importing realistic geological patterns from training images. A grid of 149*119*15 cells with grid cell size of 40 m*40 m*1 m is considered (Figure 6.5). The

Facies type	Conceptual description	Stratigraphy	Length (m)	Width (m)	Thickness (ft)
Tidal bars	Elongated ellipses w/ upper sigmoidal cross-section	Anywhere	2000-4000	500	3-7
Tidal Sand flats	Sheets (rectangular)	Anywhere Eroded by sand bars	2000	1000	6
Estuarine sands	Sheets (rectangular)	Top of the reservoir	4000	2000	8
Transgressive lags	Sheets (rectangular)	Top of the estuarine sands	3000	1000	4

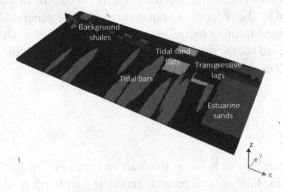

Figure 6.5 Table with rock type relations and geometrical description. The corresponding training image was generated using an unconstrained Boolean simulation method.

model includes five rock types: shale (50%), tidal sand bars (36%), tidal sand flats (1%), estuarine sands (10%) and transgressive lags (3%). Using an unconditional object-based method a training image was constructed using the following geological rules:

- Tidal sand flats should be eroded by sand bars.

- Transgressive sands should always appear on top of estuarine sands.

- An object parameterization for each facies type, except the background shale, is given in Figure 6.5.

In addition to these geological rules and patterns, trend information is available from well-logging and seismic data. Trend information is usually not incorporated in the training image itself but input as additional constraints for generating an Earth model. Since in sequential simulation an Earth model is built cell-by-cell, the trend information can be enforced into the resulting Earth model directly, no iteration is required. The training image needs only reflect the fundamental rules of deposition and erosion (the geological concept) and needs not be constrained to any specific data (well, vertical and aerial proportion variations, and seismic data). In geostatistical jargon, the training image contains the nonlocations specific stationary information (patterns, rules), while reservoir specific data enforces the trend. In this example, the following trend information was considered:

- Because of coastal influence, sand bars and flats are expected to prevail in the south-east part of the domain.

- Shale is dominant at the bottom of the model, followed by sand bars and flats, whereas estuarine sands and transgressive lags prevail in the top part.

Using the point data obtained from 140 wells, an aerial proportion map and a vertical proportion curve are estimated for each category. Figure 6.6 shows a single Earth model generated using this approach. The model matches the imposed erosion rules, depicted by the training image of Figure 6.6, matches exactly all data from 140 wells and follows the trends described by the proportion map and curves. The generated geometries are not as crisp as the Boolean training image model; there are some limitations as to what this technique can achieved but the overall joint capability of matching the data and producing realistic looking 3D models makes this an appealing technique.

6.4 Variogram-Based Methods

6.4.1 Introduction

While generating Earth models using Boolean and training image-based techniques is relatively easily to explain and understand, generating an Earth model with a variogram is less straight forward and requires a lot theory that will not be covered here. The aim,

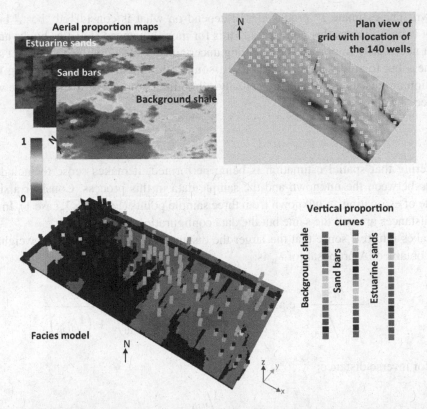

Figure 6.6 Aerial proportion map and vertical proportion curves, a single simulated facies model constrained to data from 140 wells and reflecting the structure of the training image shown in Figure 6.5.

however, is similar: to create an Earth model such that when the variogram of the cell values generated is calculated, we get back approximately the variogram that was obtained from data or other information. Evidently, there are many such models that can be created. To get at least some insight in how this is done it is necessary to discuss the theory of linear (spatial) estimation first.

6.4.2 Linear Estimation

In linear estimation the goal is to estimate some unknown quantity as a linear combination of any data available on that quantity. We use weights λ_α

$$z^*(\mathbf{u}) = \sum_{\alpha=1}^{n} \lambda_\alpha z(\mathbf{u}_\alpha)$$

$\mathbf{u}$ and $\mathbf{u}_\alpha$ denote locations in space. If this is applied to spatial problems then it is termed spatial estimation. The goal of estimation is to produce a single best guess of this unknown. The estimate z^* produced will depend on what is considered "best," but we will not further discuss this notion as (1) it is for more advanced statistics books and (2) it is not necessarily important for modeling uncertainty, whose goal is not to get a single best guess but to study alternatives. Firstly, some "old" methods for performing linear estimation are discussed, which are actually suboptimal compared to a technique which is termed "kriging."

6.4.3 Inverse Square Distance

Considering that spatial estimation is being performed, it makes sense to include the *distance* between the unknown and the sample data in this process. Consider a simple example of estimating an unknown from three sample points (Figure 6.7, case 1). In Case 2, the distances are still the same but the data configuration has changed.

It makes intuitive sense that the larger the distance taken the smaller the weight that will be obtained. A proposal for λ_α is:

$$\lambda_i = \frac{1/h_{i0}^2}{\sum\limits_{j=1}^{3} 1/h_{j0}^2} \quad \text{e.g.} \quad \lambda_3 = \frac{1/h_{30}^2}{\left(1/h_{10}\right)^2 + \left(1/h_{20}\right)^2 + \left(1/h_{30}\right)^2}$$

Or, for inverse distance:

$$\lambda_i = \frac{1/h_{i0}}{\sum\limits_{j=1}^{3} 1/h_{j0}}$$

The Euclidean distance h does not reveal much about the geology itself, it is just a distance, recall Figure 5.7. A simple improvement would be, therefore, to account for geological distance as follows:

$$\lambda_i = \frac{1/\gamma\,(h_{i0})}{\sum\limits_{j=1}^{3} 1/\gamma\,(h_{j0})}$$

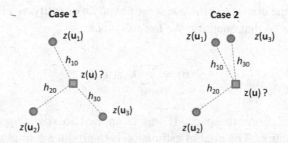

Figure 6.7　Two estimation cases with different data configurations.

However, even with this modified distance, the inverse distance is not the best technique. Consider a modified situation in Figure 6.7 (case 2). If either situation (data configuration) could be chosen for estimating the unknown, which one would be preferred? While intuitively case 1 appears more favorable, since samples are more evenly spread out, inverse distance methods give the same weight to samples 1 and 2 in both cases. This is a problem, since it is known that the information shared by sample 1 and 2 in case 2 are redundant towards estimating the unknown. Indeed, if two samples are close together, then it would be expected that they are highly correlated. Hence, one sample might be enough to contribute information towards estimating the unknown. Inverse distance methods do not take into account the *redundancy* of information. In the next section, Kriging is introduced as a technique that accounts for this redundancy when assigning weights.

6.4.4 Ordinary Kriging

Ordinary Kriging is a spatial interpolation technique that fixes many of the problems of inverse distance estimation. There is nothing "ordinary" about Ordinary Kriging. It is just a name to distinguish it from the many other forms of Kriging. In this section, the technical mathematical details of Kriging will *not* be discussed. Rather, an attempt is made to outline the properties of Kriging and learn exactly what it does. These properties can be summarized in one statement:

Kriging is the "best" linear, unbiased estimator accounting for the correlation between the data and the unknown and the redundancy of information carried by the data.

"best" can mean a lot of things. In fact, it is necessary to decide what "best" is. In Kriging, the goal is to estimate the unknowns at all locations that have not been sampled. Ideally, the estimates should be as close as possible to the true unknown values.

Kriging is a linear estimator, such that the average *squared* error between the true value and the estimate is as small as possible. For example, if the inverse distance method was applied, then it would be found that the *average squared error* is larger. The average here is taken over all unsampled locations.

Moreover, Kriging provides an unbiased estimate. That is, if Kriging is repeated a large number of times, then on average, the errors that are made will be close to zero. It makes sense that Kriging makes use of the correlation between the data and the unknown value that one want to estimate. Kriging, however, also accounts for the correlations in the data.

To find the weights, λ_α, which are needed to calculate the estimate, Kriging methods essentially solve a linear system of equations. For a simple problem with three data values, as in Figure 6.7, the system looks as follows:

$$\begin{bmatrix} \text{Var}(Z) & C(h_{12}) & C(h_{13}) \\ C(h_{12}) & \text{Var}(Z) & C(h_{23}) \\ C(h_{13}) & C(h_{23}) & \text{Var}(Z) \end{bmatrix} \begin{bmatrix} \lambda_1 \\ \lambda_2 \\ \lambda_3 \end{bmatrix} = \begin{bmatrix} C(h_{01}) \\ C(h_{02}) \\ C(h_{03}) \end{bmatrix}$$

This Kriging matrix is also termed the redundancy matrix, since it measures the redundancy between data points. Recall that C() is the covariance function, defined in chapter 5.

6.4.5 The Kriging Variance

Every estimation method *makes errors*. A best guess is never equal to the true value, unless the true value is sampled. This error has to be lived with, but at least we would like to have an idea of how much error *on average* is being made. The Kriging variance provides an idea of the magnitude of the error. In essence, if the estimation study was repeated many times, then the Kriging variance would estimate the variance of the difference between true and estimated values. Without much ado we simply list the equation emanating from theory for the ordinary Kriging variance:

$$\sigma_{OK}^2 = \text{Var}(Z) - \sum_{\alpha=1}^{n} \lambda_\alpha C(h_{0\alpha})$$

6.4.6 Sequential Gaussian Simulation

6.4.6.1 Kriging to Create a Model of Uncertainty

The goal of Kriging is to provide a single best guess for the value at an unsampled location. This is at all not the goal in this book, which is to model uncertainty. Uncertainty requires providing multiple alternatives to the true unknown value or to provide a probability distribution that reflects the lack of knowledge of that truth. This probability distribution needs to be conditioned to the available data. Recall that in the 3D training image approach this conditional probability distribution was lifted directly from the 3D training image. Hence, this probability depends on the spatial variation seen in the 3D training image. In variogram-based Earth modeling Kriging can be used to derive such probability distribution as follows.

Firstly, it is necessary to assume that a variogram or covariance model of the spatial variable being studied is provided. Assume (and this is a considerable assumption) that the conditional distribution about any unsampled location is a Gaussian/normal distribution function. Then, if it is known what the mean is of that Gaussian distribution and the variance, then we have a model of uncertainty for the unsampled value in terms of that (Gaussian) conditional distribution. A good candidate for determining this mean is the Kriging estimate, since as a best guess it provides what can be expected "on average" at this location. The Kriging variance is a good candidate for informing the variation around this "on average" value. Note that the Kriging weights depend on the variogram (or covariance), so the model of spatial continuity has been included in our uncertainty analysis. Once it is known how to determine this conditional Gaussian distribution, it is possible to proceed with sequential simulation as outlined above. This technique is then termed "sequential Gaussian simulation."

6.4.6.2 Using Kriging to Perform (Sequential) Gaussian Simulation

In order to perform sequential Gaussian simulation it is necessary to assume that all distributions are standard Gaussian. This also means that the marginal distribution, or in other words, the histogram of the variable is (standard) Gaussian. This is rarely the case; most samples obtained from the field are not Gaussian. To overcome this issue, a transformation of the variable into a Gaussian variable is performed prior to stochastic simulation. In Chapter 2 such a technique for data transformation was discussed. Then, when the simulation is finished, a back transformation is performed, which does the exact opposite of the first transformation. The complete sequential Gaussian simulation algorithm can now be summarized as follows:

1 Transform any sample (hard) data to a standard Gaussian distribution.

2 Assign the data to the grid.

3 Define a random path that loops over all the grid cells.

4 For each grid cell:

a Determine by Kriging the weights assigned to each neighboring data value or previously simulated value.

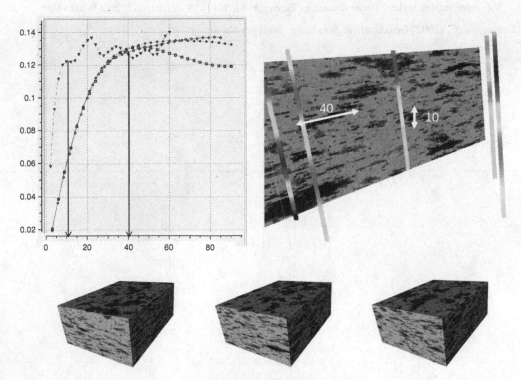

Figure 6.8 Three 3D Earth models generated using sequential Gaussian simulation (bottom). The variogram calculated from one model for two horizontal and the vertical directions (top).

b Determine the Gaussian distribution with as mean the Kriging mean and as variance the Kriging variance.

c Draw a value of that distribution.

5 Back transform all the values into the original distribution.

Figure 6.8 shows an example of sequential simulation. The input variogram used is isotropic in the horizontal direction with a range of 40 grid cells, the vertical direction has a range of 10 grid cells; there is no nugget effect. Samples of the variable (porosity in this case) are available along wells. The three resulting Earth models reflect this variogram as well as being constrained to the sample data values.

Further Reading

Chiles, J.P. and Delfiner, P. (1999) *Geostatistics: Modeling Spatial Uncertainty*, John Wiley & Sons, Inc.

Daly, C. and Caers, J. (2010) Multiple-point geostatistics: an introductory overview. *First Break*, **28**, 39–47.

Hu, L.Y. and Chugunova, T. (2008) Multiple-point geostatistics for modeling subsurface heterogeneity: A comprehensive review. *Water Resources Research*, **44**, W11413. doi:10.1029/2008WR006993.

Lantuejoul, C. (2002) *Geostatistical Simulation*, Springer Verlag.

7

Constraining Spatial Models of Uncertainty with Data

A common problem in building Earth models and constraining models of uncertainty lies in combining data sources that are indirect and at a different scale from the modeling scale with data that provide more direct information, such as those obtained through sampling.

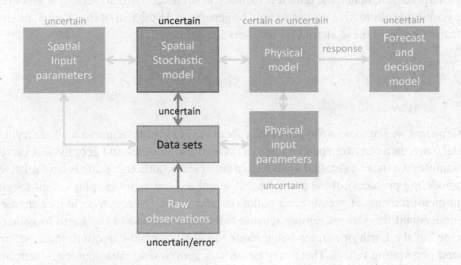

7.1 Data Integration

Data integration refers to the notion that many different sources of information are available for modeling a property or variable of interest. The question then is on how to combine these sources of information to model the spatial variable of interest. Ideally, the more data we have, the smaller the uncertainty on the variable of interest. The latter depends on how much information each data source carries about the unknown and how

Modeling Uncertainty in the Earth Sciences, First Edition. Jef Caers.
© 2011 John Wiley & Sons, Ltd. Published 2011 by John Wiley & Sons, Ltd.

redundant this source of information is with respect to other data sources in determining this unknown. In this book, two types of information have so far been dealt with: (1) hard data or direct measurements of the variable of interest at the scale at which modeling takes place and (2) spatial continuity information, that is, information on the style of spatial distribution of the property of interest as modeled in a variogram, Boolean model or 3D training image model. In this chapter all other information sources are considered. Common examples in Earth modeling are remote sensing data or geophysical measurements. A common problem in building Earth models and constraining models of uncertainty lies in combining data sources that are indirect and at a different scale than the modeling scale with data that provide more direct information, such as those obtained through sampling. Two approaches are discussed: (1) a probabilistic approach, which is relatively straightforward and easy to apply but may neglect certain aspects of the relationship between what is modeled and the data, and (2) an inverse modeling approach, which can include more information but may often be too CPU demanding or basically "overkill" for the decision problem we are trying to solve. In this chapter, we consider that "raw measurements" have been converted into "datasets" suitable for modeling. It should be understood that such a conversion process may be subject to a great deal of processing and interpretation, hence the data sets themselves are uncertain. To represent this uncertainty, multiple alternative data sets could be generated. In the rest of the chapter, methods to deal with one of these alternative data sets are discussed.

7.2 Probability-Based Approaches

7.2.1 Introduction

Assume that we have various types of data about the 3D Earth phenomena we are trying to model, such data can often be classified in two groups: samples and geophysical surveys. By samples we mean a detailed analysis at a small scale (although scale is here relative to the modeling problem); these could be soil samples, core samples, plugs, well-logging, point measurements of pressure, air pollutions and so on. By geophysical measurements we understand the various remote sensing techniques applied to the Earth to gather an "image" of the Earth or surface being modeling (e.g., synthetic aperture radar, seismic, ground penetrating radar). There may be various geophysical data sources (electromagnetic, seismic, gravity) and various point sources. Other type of measurements may be available that are indicators of what is being modeled: in general such information is named "soft information." In this section, the following are addressed:

- How to use data sources such as geophysical measurements or "soft information" in general to reduce uncertainty about what is being modeled (and hopefully about decisions being made, see also Chapter 11).

- How to account for the "partial information" that such data sources provide.

- How to combine several data sources (e.g., several geophysical sources, or point samples with geophysical sources) each one of which provides only partial information.

A data source often provides only partial information about what we are trying to model. For example, seismic data (Chapter 8) does not provide a measurement of porosity or permeability, properties important to flow in porous media; instead, it provides measurements that are indicators of the level of porosity. Satellite data in climate modeling do not provide direct, exact information on temperature, only indicators of temperature changes. Given these criteria, we will proceed in two steps in order to include these data in our model of uncertainty:

1 **Calibration step:** how much information is contained in each data source? Or, what is the "information content" of each data source?

2 **Integration step:** how do we combine these various sources of information content into a single model of uncertainty?

7.2.2 Calibration of Information Content

The amount of information contained in a data source is dependent on many factors, such as: the measurement configuration, the measurement error, the physics of the measurement, the scale of modeling, and so on.

The first question that should be asked is: how do we model quantitatively the information content of a data source. In probabilistic methods a conditional distribution is used (Chapter 2). Recall that a conditional distribution $P(A|B)$ models the uncertainty of some target variable A, given some information B. In our case B will be the data source and A will be what is being modeled. Recall also that if $P(A|B) = P(A)$ then B carries no information on A. The question now is how is $P(A|B)$ determined?

To determine such conditional probability, more information is needed, more specifically, we need data pairs (a_i, b_i), that is, mutual or joint observations of what we are trying to model and the data source. This means that at some limited set of locations it is necessary to have observed the true Earth as well as the data source. In many applications, at the sample locations, it is possible to have information on A as well as B.

For example, from wells, there may be measurements of porosity and from seismic measurements of seismic impedance (or any other seismic attribute) in 3D (such as shown in Figure 7.1). This provides pairs of porosity and impedance measurements that can be used to plot a scatter plot, such as shown in Figure 7.2. In this scatter plot $P(A|B)$ can now be calculated, where the event "$A = $ (porosity $< t$)", for some threshold t, and "$B = (s <$ impedance $< s + \Delta s)$", as shown in Figure 7.2. In this way a new function is created:

$$P(A|B) = \varphi(t, s)$$

Once we have this function, the conditional probability for any t and any s can be evaluated. This function is a "calibration function" that measures how much information impedance carries about porosity. There are many other ways to get this function. Physical approaches such as rock physics may provide this function, or one may opt for statistical techniques (e.g., regression methods such neural networks) to "lift" this function from a data set belonging to another field if what occurs in that field is deemed similar.

Figure 7.1 Calibration data set: samples providing detailed but only local information (left); a geophysical image of the earth providing a fuzzy but global insight (right). At the sample locations we have both sets of observations.

7.2.3 Integrating Information Content

Consider now the situation where several such calibrations have been obtained because many data sources are available. In other words, several $P(A|B_1), P(A|B_2), \ldots$ have been obtained. The next question is: how do we combine information from individual sources

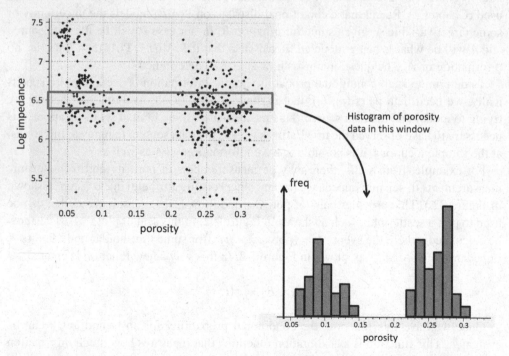

Figure 7.2 Calibration of porosity from seismic using a scatter plot. By considering porosity within a given window of seismic impedance, it is possible to calculate the frequencies of porosity that are less than a certain threshold value from the histogram shown.

as expressed through these conditional probabilities into a single model probability based on all sources, namely:

$$\text{What is } P(A|B_1, B_2, \ldots)?$$

One way is to perform a single calibration involving all the data sources at once to get this combined conditional probability, but this is often too difficult or it would require a high quality and rich calibration data set. A single calibration with many variables requires a lot of data in order for this calibration to be accurate. Often such exhaustive data sets are not available. Also, these partial conditional probabilities may be provided by experts from very different fields. In climate modeling, for example, one may have very different data sources, such as tree ring growth, ice cores, pollen and sea floor sediment cores, to predict climate changes, each requiring a very different field of expertise. It would be too difficult to put all data in one basket and then hope to directly get a good prediction of climate change.

In other words, some way to combine these individual conditional probabilities into a joint conditional probability is needed. A simple and quite general way of doing this combination of probabilities is provided here, knowing that other methods exists in the literature, but typically they require making similar assumptions. To understand better the issues in doing this, consider a very simple binary problem: two sources of information (events B_1 and B_2) inform that the chance it will rain tomorrow (event A) is significant, in fact:

- From the first source B_1 we have deduced (by calibration for example) that there is a probability of 0.7 of event A occurring.

- From the second source B_2 we obtain a probability of 0.6.

- The "historical" probability of it raining on the date "tomorrow" is 0.25.

- We know that the two sources B_1 and B_2 do not encapsulate the same data (calibration data or experts opinion).

The question is simple: what would you give as the probability for it to rain tomorrow? The answer to this question is not unique and depends how much "overlap" there is in the information of each source in determining the event A. Clearly, if the two sources (e.g., experts, calibration data) use the same data to come to their respective conditional probabilities then there is a conflict. This is often the case in practice, since no two procedures for modeling conditional probabilities need to yield the exact same result because of various modeling and measurement errors. But we will assume here that at least theoretically we do not have such conflict.

What is also relevant is that the "prior" or "background" conditional probability is 0.25. In fact, what is very relevant is that both sources predict a higher than usual probability of it raining. To capture this amount of "overlap" a new term is introduced, namely that of data redundancy. Redundancy measures how much "overlap" there is in the sources of information in terms of predicting an event. If there is complete overlap then the information sources are redundant (and you can throw one away). Redundancy should not be confused with dependency. Redundancy is always with regard to a target event (A = "raining tomorrow"). Dependency is the degree of association between information sources B_1 and B_2 regardless of the target A. As a consequence, redundancy will change when the target A changes. Sometimes information is completely redundant. For example, if you buy 100 copies of the same New York Times (100 information sources), then you will not get more information about a news item than buying the paper just once. However, buying two different newspapers to get information about one story may provide more insight then just buying one (if they do not share a common information source of course) If you let 100 experts look independently at some data to draw a conclusion about some phenomenon, then they will be less redundant than when you put all the experts in the same room, where they may try to agree on what is happening, thereby influencing each other, or resort to "herding," that is, following the voice of a single expert and therefore be more redundant.

Redundancy is difficult to "measure" or quantify, it requires making assumptions about the nature of the information sources with respect to the target unknown. If it is difficult to determine, we can start by making an assumption or hypothesis and then check if the results we get are reasonable. One such assumption that can be made is as follows:

> The relative contribution of information B_2 to knowledge of A is the same regardless of the fact that you have information B_1

This statement indeed contains some assumption on redundancy, that is, how we use B_2 to determine A is the same whether we know B_1 or not (it does not say we are not using B_1!). What follows is then a quantification of this statement in terms of probabilities. Consider firstly the following quantities derived from conditional probabilities:

$$b_1 = \frac{1 - P(A|B_1)}{P(A|B_1)}, \quad b_2 = \frac{1 - P(A|B_2)}{P(A|B_2)}, \quad a = \frac{1 - P(A)}{P(A)}$$

Each such scalar value can be seen as a distance, namely if $b_1 = 0$ then $P(A|B_1) = 1$, hence B_1 tells everything about A, while if $b_1 =$ infinite then B_1 makes sure that A will not happen. What we would like to know is:

$$x = \frac{1 - P(A|B_1, B_2)}{P(A|B_1, B_2)} \Rightarrow P(A|B_1, B_2) = \frac{1}{1 + x}$$

If we look closely at our hypothesis, then we can equate the following:

$$\frac{b_2 - a}{a} = \text{the relative contribution of information source } B_2 \text{ when not having } B_1$$

Notice how this is a relative contribution, namely relative to what is known before knowing the information B_1. Before knowing B_1 we have prior information $P(A)$ or in terms of distance, simply a.

$$\frac{x - b_1}{b_1} = \text{the relative contribution of information source } B_2 \text{ when } B_1 \text{ is available}$$

The relative contribution is now relative to knowing B_1. Hence, our hypothesis can be formulated into an equation as follows:

$$\frac{b_2 - a}{a} = \frac{x - b_1}{b_1} \Rightarrow x = \frac{b_1 b_2}{a} \quad \text{or} \quad P(A|B_1, B_2) = \frac{a}{a + b_1 b_2}$$

If we return to our example, then we can use the above equations to calculate

$$P(A|B_1, B_2) = 0.91$$

which suggests that we have more certainty about it raining from these two data sources than from considering each data source separately. In other words, the model assumption enforces "compounding" (or re-enforcement) of information from the two sources.

7.2.4 Application to Modeling Spatial Uncertainty

Next consider how to use this model of data redundancy in the context of spatial uncertainty. Recall that spatial uncertainty was represented by generating many Earth models that reflect a spatial model of continuity. Using the above ideas a new set of alternative Earth models that are in addition to the spatial continuity model also reflective of other data sources can now be built. Consider the example of geophysical data and consider determining the presence or absence of sand at each grid block in an Earth model. Consider that we have determined the function $P(A = \text{sand}|B_2)$; B_2 is the geophysical information. This means that at every location in the grid where we have a geophysical measurement, this measurement can be replaced with a conditional probability by means of the calibration function. Figure 7.3 shows the actual case, the color in the middle plot represents this probability of sand for the given geophysical information (in this case, seismic data, which is not shown). Next to this data, is the spatial continuity as expressed in the training image on the left. What we are trying to accomplish here, is to create an Earth model that has more sand channels where the probability of sand is higher. This is simply done by means of an extension of the sequential simulation methodology.

Recall the sequential simulation algorithm:

1 Assign any sample (hard) data to the simulation grid.

2 Define a random path that loops over all the grid cells.

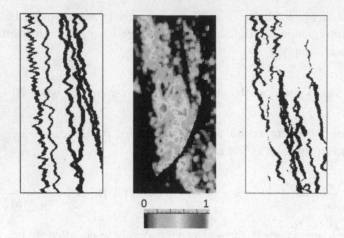

Figure 7.3 Training image (left), probability derived from seismic (middle), Earth model (right).

3 At each grid cell:

a Determine the uncertainty of that unsampled grid value given the data and the previously simulated cells values in terms of a probability distribution.

b Sample a value from this probability distribution by Monte Carlo simulation and assign it to the grid.

Step a in this algorithm consists now of three steps:

1 Determine from the 3D training image $P(A|B_1)$ by scanning.

2 Determine from the sand probability cube (at that grid cell location) the probability $P(A|B_2)$.

3 Use the above equations to determine $P(A|B_1, B_2)$ (note that $P(A)$ is the proportion of sand).

In step b a Monte Carlo simulation is performed using $P(A| B_1, B_2)$. Figure 7.3 (right) shows an example Earth model created with this algorithm.

7.3 Variogram-Based Approaches

When dealing with continuous properties that do not have a very distinct spatial variation, it was discussed in Chapter 5 that variogram-based models of spatial continuity could be an appropriate choice. Indeed, the 3D training image and Boolean models of spatial continuity assume that there is a distinct and specific spatial variation present that needs to be modeled, either through objects or through a conceptual 3D image.

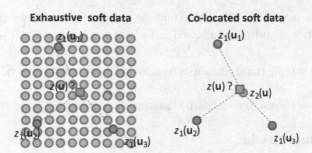

Figure 7.4   Two configurations of soft data: exhaustive and co-located.

The question is now how to constrain these variogram-based Earth models to additional data sources such as, for example, the 3D geophysical image of Figure 7.1. Recall that variogram models are based on the degree of correlation between data values. Indeed, we actually started the discussion on variograms with the correlogram, and derived such correlogram from scatter plots and the calculation of correlation coefficient between scatter plots. To additionally constrain the models to another data source we will be relying on the same principle. In traditional geostatistics, this other data source is also termed "soft data."

Consider a configuration of data similar to that discussed when introducing the notion of Kriging (Figure 7.4). However, in addition "soft data" are available which are typically exhaustive, as would be the case of geophysical or remote sensing data. Accounting for all this data is difficult, so the situation is often simplified by retaining only the co-located (or at least a much more limited amount) data value. If the soft data are indeed much more smoothly varying, see for example Figures 7.1 and 7.3, then this is a reasonable assumption. The smooth variation in the soft data makes such data redundant towards determining the unknown value. We can now simply extend linear estimation to include this additional data value (note that this data value may have a different unit of measurement than the hard data):

$$z_1^*(\mathbf{u}) = \sum_{\alpha=1}^{n} \lambda_\alpha z_1(\mathbf{u}_\alpha) + \lambda_{n+1} z_2(\mathbf{u})$$

This additional weight will depend on the degree of correlation between the value of the soft data z_2 measured at $\mathbf{u}$ and the variable being modeled. Such information can be obtained simply by creating a scatter plot such as in Figure 7.2 and calculating the correlation coefficient between the "hard variable" z_1 and the "soft variable z_2." This makes sense intuitively: if that correlation were to be zero, than the weight λ_{n+1} should be zero because the soft data is uninformative, while if that correlation is one, then the soft data gets all the weight, while the rest gets assigned zero weights. The actual derivation and calculation of these weights is left out here; it is a topic for geostatistics books. What is important to remember is that, in addition to the variogram model, one needs to know additionally at least the correlation coefficient.

Once we know how to use Kriging to include the additional data source the sequential Gaussian simulation algorithm is adapted as follows:

1 Transform the sample (hard) data to a standard Gaussian distribution.

2 Transform the soft data to a standard Gaussian distribution.

3 Assign the data to the grid.

4 Define a random path that loops over all the grid cells.

5 For each grid cell:

a Determine by Kriging the weights assigned to each neighboring data value or previously simulated value, including the weight associated to the soft data.

b Determine the Gaussian distribution with as mean the Kriging mean and as variance the Kriging variance.

c Draw a value of that distribution.

6 Back-transform all the values into the original distribution function.

Figure 7.5 shows an example. We have five wells with measurements of porosity (Figure 7.1). The scatter plot is shown and a correlation coefficient of -0.49 is calculated. Next this correlation coefficient is included in the sequential simulation algorithm and one Earth model generated (Figure 7.5). If the correlation coefficient is altered to -0.85 then another Earth model is obtained (Figure 7.5) that share many more features with the soft data (where the high and low values occur) than using a lower correlation.

7.4 Inverse Modeling Approaches

7.4.1 Introduction

In the above probabilistic approach the information content of each data source is captured with a conditional probability as modeled from a calibration data set. This approach is quite useful to get a model of uncertainty quickly, for the given data set, if such calibration data sets are reliable and informative about the relationship between the variable being modeled (e.g., temperature) and the data source (e.g., satellite data). Sometimes a calibration data set may not be available, or such data may provide incomplete information about that relationship. Consider a simple example. Suppose we have a calibration data set such as in Figure 7.6, the variable being modeled is A while the data source is B. If the probabilistic methods are employed, such as in Figure 7.2, then in such methods we tend to assume that the relationship is linear, there is no other reason for assuming

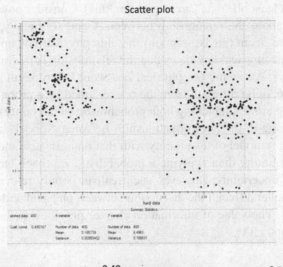

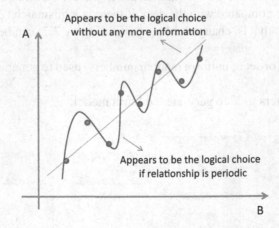

Figure 7.5   Example of sequential simulation with soft data.

Figure 7.6   Two ways of modeling the relationship between A and B.

anything else (or at least no "data" are suggesting this). Consider now that we have some more information about the relationship between *A* and *B*, namely, that there exists a physical law and we deem (an assumption) that this physical law applies to variables *A* and *B*; moreover, the physical law states that the relationship is periodic (oscillating with a certain frequency). Assuming the physical law would then result in a relationship as shown in Figure 7.6, quite different from the one assumed by a probabilistic approach.

Inverse modeling aims at including both probabilistic (uncertainty) as well as physical (deemed deterministic in the book) relationships between "model" and "data" with the aim of constraining a model of uncertainty with that data. Inverse modeling is more difficult and time consuming than applying a probabilistic approach but may lead to more realistic models of uncertainty. However, such realism is only relevant to the particular decision problem at hand (recall the discussion between physical and deterministic modeling in Chapter 3). The value of information chapter provides methods for making this judgment call (Chapter 11).

7.4.2 The Role of Bayes' Rule in Inverse Model Solutions

Figure 7.7 describes the various elements involved (refer also to Figure 3.2) in inverse modeling. A spatial model containing all relevant variables is built; this requires spatial input parameters (Chapters 5 and 6). The data are considered to be a "physical response" (a seismic signal, a pressure pulse, a temperature change over time) emanating from the Earth because of some testing applied to it. For example, we can hit the ground with our boot at location X and measure the vibrations over time at location Y. These vibrations provide data over some region of the subsurface between X and Y. A forward model consists of modeling (through partial differential equations for example) or simulating (on a computer) how these vibrations travel from source X to location Y for an assumed spatial variation of all properties in the Earth (as modeled in an Earth model for example). The response obtained from the forward model is the forward response. This forward response can then be compared with the data. If there is a mismatch (Δ), then something needs to be adjusted, that is, changed. Many things (Figure 7.7) can be changed namely:

1 The random seed (or set of uniform random numbers) used to generate the Earth model.

2 The input parameters used to generate the Earth model.

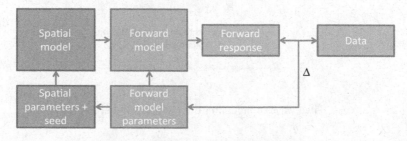

Figure 7.7   Elements of inverse modeling.

3 The physical parameters related to the forward model/response.

4 The initial and boundary condition related to the forward model/response.

5 The physical law(s) itself on which the forward model is based.

6 The data may not be correct or prone to error, so it may not be necessary to match data exactly.

One can iterate many times, that is, changes can be applied to the Earth model, the forward response calculated and the mismatch updated. In doing so, two questions are should be considered:

1 What do we change?

2 How do we change things (parameters, conditions, laws)?

There is no unique method that will address these questions exactly; often this is based on judgment and the level of expertise in this area of modeling and often requires a combination of techniques. Question 1 has already been addressed somewhat in this book (Chapter 4). In addressing question 1, two guidelines should be considered:

1 We should change that which impacts matching the data better.

2 We should change that which impacts the decision question for which the models are constructed

The traditional view in inverse modeling is to consider only question 1, that is, establishing what parameter or variable matters to matching the data. However, it may not be necessary to match the data exactly if that does not impact the decision question, or if the variable we are changing does not impact the decision question. Chapter 11 on value of information provides some ideas on how to address obtaining a balance between these two questions.

To figure out what is most sensitive (either to data or decision question), we also refer to Chapters 10 and 11 on response uncertainty and value of information for additional techniques in determining sensitivity. The second question has many answers, that is, there are many methods that can be applied, but each of these methods can be framed with probability theory and Bayes' rule. Before doing so, consider a typical example problem.

Consider an Earth model as $\mathbf{m}$, for example a 3D grid of possibly multiple properties. Consider the data set as $\mathbf{d}$, that is, a list of observations (such as a time series of amplitudes that represent the above mentioned vibrations); take that the observations are exact, that is, no measurement error. If we take $A = \mathbf{M}$ and $B = \mathbf{D}$ (capital for random variables bold to denote vectors), then Bayes' rule simply becomes:

$$P(\mathbf{M} = \mathbf{m} | \mathbf{D} = \mathbf{d}) = \frac{P(\mathbf{D} = \mathbf{d} | \mathbf{M} = \mathbf{m})\, P(\mathbf{M} = \mathbf{m})}{P(\mathbf{D} = \mathbf{d})}$$

A synthetic (i.e., constructed) example that will be further explored is shown in Figure 7.8. A pumping test is applied to a synthetically generated field of hydraulic conductivity (a measure of how well each fluid flows in a porous medium) that has a binary type of spatial variation (channels vs background), the channels have higher conductivity than the background and these values are known. What is not known is the location of the channels, we only know the "style" of channeling as expressed in the training image

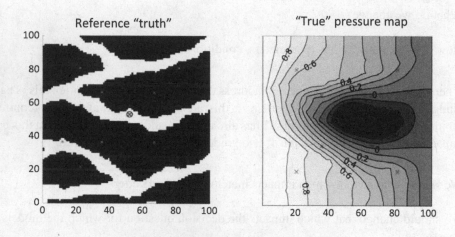

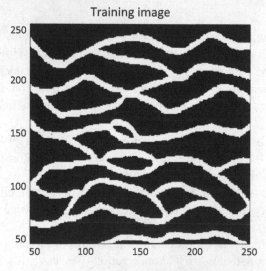

Figure 7.8   Case study for inverse modeling.

below (we assume this also known, although in most cases it would be uncertain as well). **M** is therefore a vector of 100×100 binary variables (1/0), for each grid-block indicating whether or not a channel is present. The pump test at the central well location in Figure 7.8 is done such that the pressure in the surrounding area will decrease. To simulate this pressure decrease a "pressure simulator," based on the physics of flow in porous media, is employed. This gives nine pressure readings at the nine locations marked by red crosses in Figure 7.8; hence **d** is a nine-dimensional vector containing these nine pressure readings. The question is now to find all the possible variations of spatial distribution of channels that match these nine data.

To explain the role of Bayes' rule, consider first a simpler example, where m and d are discrete and univariate variables. This allows easy modeling of probabilities. The above example is returned to later. Consider that d has been measured and has value $d = 1$. m has 10 possible outcome $0, 10, 20, \ldots, 100$; however, even prior to considering any data it is known that some outcomes m are more probable than others. For example, m can be the proportion of sand in a deltaic aquifer system. Around the world, one may have examined many aquifer systems similar to the one being considered and from such examination a table of "prior" frequencies has been tabulated as follows:

m	Prior probability $P(M = m)$
0	0
10	5
20	20
30	30
40	25
50	10
60	10
>60	0 in %

Suppose d is some physical measurement (units don't matter) and we have a physical law that provides the forward response of g as applied to m as follows

m	g(m)
0	1
10	1
20	1
30	2
40	2
50	2
60	3
>60	4

It is clear that if $d = 1$ then only three solutions are possible, $m = 0$, 10, and 20. What is the probability for each solution? Are they equal? If Bayes' rule is considered, then this depends on the prior probabilities for those three solutions namely $P(M = m) = 0, 5$ and 20% as well as the prior for d. We do not have the prior for d, but we don't need it, since we know that probabilities $P(M = m|D = d)$ for all m must add up to one, hence we can simply re-standardize 5% and 20% ($M = 0$ cannot occur according to the prior) to sum up to one and obtain:

$$P(M = 10|D = 1) = 20\%$$
$$P(M = 20|D = 1) = 80\%$$

$P(D)$ could have been determined as follows (Chapter 2):

$$P(D = d) = \sum_{m=0}^{100} P(D = d|M = m)\,P(M = m)$$

Note that in this case $P(D = d |M = m)$ is one if $m = 0$, 10 and 20% and zero in all other possibilitie:, so

$$P(D = 1) = 100\% \times 0\% + 100\% \times 5\% + 100\% \times 20\% = 25\%$$

Suppose now that there is measurement error. What exactly is a "measurement error"? In probabilistic terms it means that if we would repeat the same measurement many times, we would get different results, that is we can measure at one instance $d = 1$ but it could be incorrect; for example, if we measure again, we could have measured $d = 2$ instead. However, measuring something ($d = 1$) does not tell us what the measurement error is. How then can this error be quantified in a general way? First, it makes sense that the measurement error is function of what is being measured: some quantities (m) are easier to measure than others. For example, it may be easier to measure a large quantity than a very small quantity. To model this dependency, one can therefore use a conditional probability of the form $P(D = d|M = m)$ to express measurement error, that is, if M is a certain value m, what is then the probability of the measurement occurring? If this probability is one for a particular d then there is no error. This probability is termed a likelihood probability and can only truly be determined if the measurements for various m are repeated. In practice, often this cannot be done (too costly), so we can assume some likelihood probability, rely on past experience or try and make models of errors (for both measurement and the model g), which are not covered in this book. Suppose that these likelihoods have been determined and are:

$$P(D = 1|M = 10) = 75\% \text{ and } P(D = 1|M = 20) = 25\%$$

If Bayes' rule is applied, again in the case that $D = 1$ is measured, we obtain:

$$P(M = 10|D = 1) = \frac{75 \times 5}{75 \times 5 + 25 \times 20} = 0.428$$

$$P(M = 20|D = 1) = \frac{25 \times 20}{75 \times 5 + 25 \times 20} = 0.572$$

Note that there is a risk in specifying the prior and a likelihood probability independently: it may occur that for all conditional events $M = m$ in the likelihood, all prior probabilities are zero (or just very small). This occurs often in practice.

In real Earth modeling **M** and **D** are multidimensional but the same probability rules apply. As discussed in various previous chapters, the goal is to create several Earth models as representative of a model of uncertainty. Firstly, consider generating Earth models without the data **d**. Any technique described in Chapters 6–8 can be used to generate them. We denote these Earth models as $\mathbf{m}^{(1)}, \mathbf{m}^{(2)}, \ldots, \mathbf{m}^{(L)}$ and consider them "sampled" from some prior model $P(\mathbf{M})$. However, this prior probability is not explicitly stated as done for the simple example above, it would be too difficult because of the high dimension of **m**. This set of models is termed "prior Earth models." For the case study of Figure 7.8, nine such prior Earth models $\mathbf{m}^{(j)}$ are shown in Figure 7.9 and generated with the training-image based techniques described in Chapter 6. Next, among this possibly large set (much more than nine!) it is necessary to search for Earth models that match the data

$$\mathbf{m}^{(i)} \text{ such that } g(\mathbf{m}^{(i)}) = \mathbf{d}$$

if no uncertainty nor error are assumed in the data–model (**D–M**) relationship. This new set of models represents the "posterior" distribution $P(\mathbf{M} = \mathbf{m}|D = \mathbf{d})$ or the posterior uncertainty (posterior to including data; Chapter 3). The models are termed "posterior Earth models." If measurement error occurs then no exact matching is desired (see next section for finding models in this case). However, the problem may be that (i) no such models can be found or (ii) it could be very difficult (CPU demanding) to find multiple posterior Earth models if the function g takes a long time to evaluate, such as in the case of a numerical simulator. In the first case, there is likely an inconsistency between the set of prior models that was chosen and the data–model relationship (the physical law); this is a common problem. For the second case specialized techniques are needed; these are briefly discussed at the end of this chapter.

Several conclusions can be drawn from this discussion on the role of Bayes' rule:

- If Bayes' rule and probability theory are used as a modeling framework, then any inverse solution **m** depends on the (subjective) prior probability that is stated *and* the data–model relationship (physical law), which may be prone to error or uncertainty because the data has measurement error or the physical law may not be well known. Bayes' rule is also clear that one cannot escape stating a prior probability, that is, if one doesn't know this prior probability then simply taking all probabilities equal or Gaussian for that matter is a very specific and equally subjective choice of a prior probability.

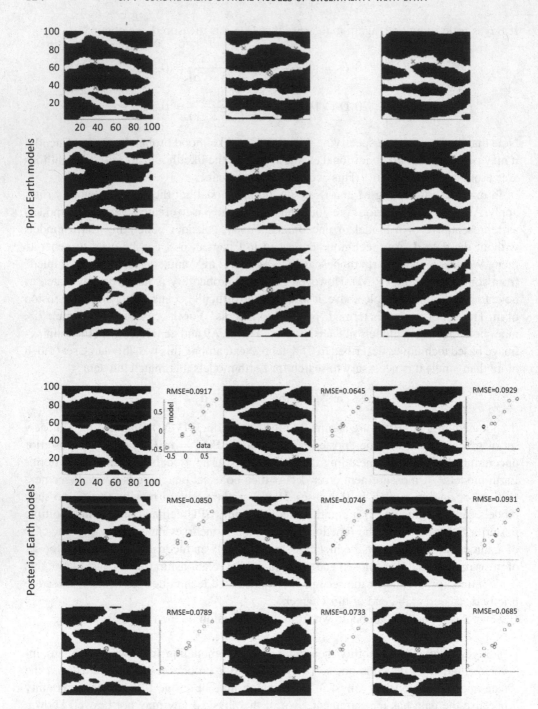

Figure 7.9 Nine prior Earth models and nine posterior Earth models. The RMSE indicates how well they match the data at the nine observation locations. The small cross-plots on the right of each Figure show the mismatch of the nine pressure measurements.

- The error in the data–model relationship is modeled as a conditional probability and this error may depend on what is being measured, that is, small things may be harder to measure than big things.

- There may be inconsistency between the prior and the data–model relationship. Often in practice one chooses these two probabilities separately; for example a geologist may be responsible for generating several prior models and a hydrologist may be responsible for specifying the physics of subsurface flow as well as determine any errors that may exist in gathering real field data. This will be common in any modeling of uncertainty that requires many disciplines. These two persons may specify two probabilities that are inconsistent, or implicitly assume that one is correct and the other is not, without any specific evidence as to why (often the data, that is, likelihood probability is chosen as correct and not the prior).

7.4.3 Sampling Methods

7.4.3.1 Rejection Sampling

Bayes' rule does not tell us how to find solutions to inverse problems, in other words, it is not a technique for finding "posterior Earth models"; it just tells us what the constraints are to finding them. A simple technique for finding inverse solutions is called the rejection sampler. In fact, rejection sampling is much in line with Popper's view on model falsification (Chapter 3), that is, we can really only "reject" those models that are proven to be false from empirical data. Although, Popper's principle is more general than a typical rejection sampler, namely, he considered everything imaginable to be uncertain, including physical laws, while in rejection sampling we typically assume only Earth models to be uncertain. Popper's idea, jointly with Bayes' rule emphasizes the important role of the prior, namely, all that can be imagined should be included in this prior, only then, with data, can we start rejecting that what can be proven false by means of data.

Consider first that we want to match the data exactly, then rejection sampling consists of doing the following:

1 Generating a (prior) Earth model **m** consistent with any prior information.

2 Evaluating $g(\mathbf{m})$.

3 If $g(\mathbf{m}) = \mathbf{d}$, then keeping that model; if not, reject it.

4 Go to step 1 until a desired amount of models has been found.

In this simple form of rejection sampling, models that do not have an exact match to the data (as defined by $g(\mathbf{m}) = \mathbf{d}$) are rejected. The Earth models accepted follow Bayes' rule, since only models created from the prior are valid models. If the prior is inconsistent with g, that is, no models can be found, then either g or the way the prior models are generated should be changed. Often, there is no unique, objective criterion to choose which one should be changed. Note that in order to generate a prior model, several sources of uncertainty could be considered: the training image, the random seed, the variogram, the mean and so on, and each of these "parameters" can be changed. In step 1 parameters are changed randomly, for example if we have uncertainty about the training image, and two possible training images have been specified, one with 35% probability and the other with 65% probability, then 35% of the Earth models should be generated using training image 1. This is the exact meaning of "Earth models are consistent with prior information." If more parameters are uncertain, then each parameter is drawn randomly at least if all parameters are considered as being independent variables (recall Chapter 2).

In most cases it is not necessary to achieve a perfect match to the data, since many errors exist in specifying the data-to-model relationship. This uncertainty is specified through the conditional probability $P(\mathbf{D}|\mathbf{M})$. In that case, the following adaptation (no proof given here) of the rejection sampler has been proven to be consistent with Bayes' rule:

1 Generate a (prior) Earth model **m** consistent with any prior information.

2 Evaluate g(**m**).

3 Accept the model **m** with probability $p = P(\mathbf{D} = \mathbf{d}|\mathbf{M} = \mathbf{m})/P^{\text{max}}$.

4 Go to step 1 until a desired amount of models has been found.

Step 3 needs a little more elaboration. P^{max} is the maximum value of the conditional probability $P(\mathbf{D} = \mathbf{d}|\mathbf{M} = \mathbf{m})$. Accepting a model with a given probability p is done as follows:

a Draw from a uniform distribution a value $p*$ (Chapter 2).

b If $p* \leq p$, then accept the model, otherwise reject it.

We apply the rejection sampler to the synthetic case study of Figure 7.8. The only unknown is the location of the channels, so there is only "spatial uncertainty," there is no "parameter uncertainty" (the training image and any other parameter are assumed known). To specify the likelihood probability proceed as follows

$$P(\mathbf{D} = \mathbf{d}|\mathbf{M} = \mathbf{m}) \simeq \exp\left(-\frac{\text{RMSE}(\mathbf{m}, \mathbf{d})^2}{2\sigma^2}\right) \quad \text{with } \sigma = 0.03 \quad \text{and}$$

$$\text{RMSE} = \sqrt{\frac{1}{9}\sum_{k=1}^{9}(g_i(\mathbf{m}) - d_i)^2}$$

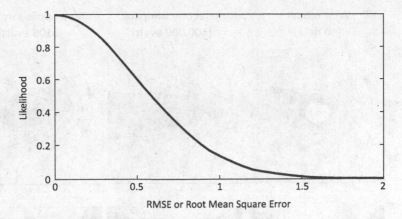

Figure 7.10   Example of a likelihood function, in this case a Gaussian likelihood; note that $P^{max} = 1$.

see also Figure 7.10. RMSE is the Root Mean Squared Error. This function, termed a Gaussian likelihood, becomes smaller when the mismatch between data and forward response is larger, the error as stated between the nine pressure data d_i and the nine forward simulated pressure responses $g_i(m)$. In Figure 7.9, nine out of 150 Earth models that match the data are shown; these are termed the posterior Earth models. To obtain these 150 Earth models, it was necessary to iterate the rejection sampler about 100 000 times, that is, 100 000 models **m** were generated and tested with the above procedure. A summary of these models is shown in Figure 7.11. To summarize a set of Earth models what is termed the ensemble average (EA) is calculated. This is the cell-wise average of all the images in a set:

$$EA = \frac{1}{L} \sum_{\ell=1}^{L} \mathbf{m}^{(\ell)}$$

Next, the ensemble standard deviation can also be calculated

$$ESD = \sqrt{\frac{1}{L-1} \sum_{\ell=1}^{L} (\mathbf{m}^{(\ell)} - EA)^2}$$

which can be regarded as some measure of local variability between the images. If one adds up images that are binary then the probability of occurrence (of channel in this case) is obtained. Figure 7.11 shows the ensemble average of the pressure maps (termed heads) for 150 prior Earth models and the 150 posterior Earth models. Clearly, the pressures are different between prior and posterior, which means that the data is informative about what is being modeled, hence the standard deviation of the posterior ensemble is smaller than that of the prior ensemble, with most of the reduction occurring near the location of the central pumping well. Remark also that the prior is not informative at all about

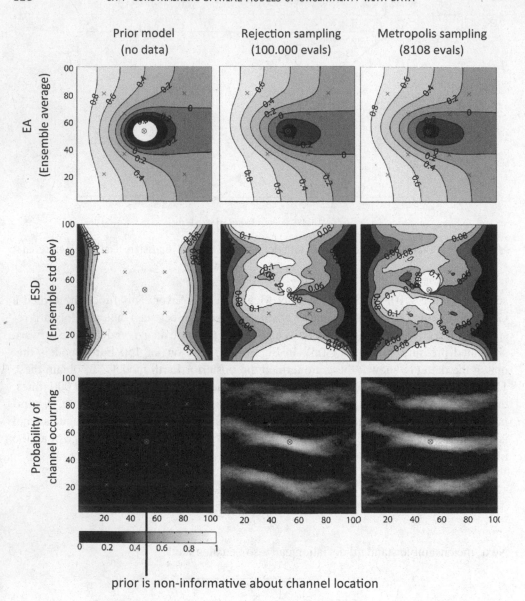

prior is non-informative about channel location

Figure 7.11 Ensemble average, standard deviation and probability of channel occurring for (a) no data used, (b) the rejections sampler and (c) the Metropolis–Hastings sampler.

the specific location of the channels (therefore the gray plot in the left-hand corner). The prior only informs the "style" of channel distribution.

7.4.3.2 Metropolis Sampler

The rejection sampler is called an "exact" sampler in the sense that it follows perfectly Bayes' rule. However, this perfection comes at a price: it may take an enormous amount of CPU time to obtain enough posterior Earth models in the ensemble to get an adequate model of uncertainty. It may be argued that following Bayes' rule exactly is not desired. This is a valid argument since many components within the framework of modeling uncertainty are indeed uncertain, such as the prior probability, the data and the physical law, and, as discussed in Chapter 3, there is no objective criterion to verify/validate that these are correct. As with any law, principle and rule, Bayes' rule leads only to "internal consistency" (Chapter 3). Bayes' rule should therefore be seen as a template or framework, not a truth, within which many problems can be solved and moreover solved consistently according to a well established probability principle. Completely abandoning this rule may (but not necessarily) lead to an artificial reduction of uncertainty as will be discussed in the next section.

A slightly different but much more efficient (less iterations needed) sampler from the rejection sampler is the Metropolis sampler. There are many forms of this sampler, but in one of its most simple form it works as follows (again, no proof is given):

1 Generate a (prior) Earth model **m** consistent with any prior information.

2 Propose a change to **m**, but make sure this change is still consistent with any prior information. Call this change **m***.

3 If $P(\mathbf{D} = \mathbf{d} \mid \mathbf{M} = \mathbf{m}) < P(\mathbf{D} = \mathbf{d} \mid \mathbf{M} = \mathbf{m}^*)$ then

$$\text{accept } \mathbf{m}^* \text{ (call } \mathbf{m} \text{ now } \mathbf{m}^*)$$

or

$$\text{accept } \mathbf{m}^* \text{ with probability } p = \frac{P(\mathbf{D} = \mathbf{d} \mid \mathbf{M} = \mathbf{m}^*)}{P(\mathbf{D} = \mathbf{d} \mid \mathbf{M} = \mathbf{m})}$$

4 Go back to step 2 and iterate many times "*until convergence.*"

5 Keep the Earth model that remains after convergence.

The mechanism here is such that models with increased likelihood probability are accepted all the time, but even models that have lower likelihood probability can still be accepted. If only models **m*** with increased likelihood were accepted then we would get trapped quickly, that is, we wouldn't find any improvement to matching the data better.

In step 2, also termed the "proposal step," one has a lot of degree of freedom and most of the practical success of this algorithm lies in finding good change mechanisms. An example "change" would simply be to create a completely new Earth model **m*** but this is often inefficient, more gradual changes are often preferred. Rather we want to make changes where it matters (as to matching the data better for example) by considering the most sensitive parameters or areas in the Earth model deemed more consequential. However, any such change should be consistent with the prior information, that is, we cannot make arbitrary changes that could that are in conflict with such prior (see the comments on the role of Bayes' rule).

Step 4 needs more elaboration. The theory states that one should iterate the above algorithm an infinite number of times to get an Earth model that follows Bayes' rule and, hence, be equivalent to the rejection sampler. This is evidently not practical, since the goal here is to obtain models relatively fast. Instead, we iterate till we do not observe much change in some statistical summaries of the Earth models found.

Applying the Metropolis algorithm to our example (we won't elaborate on the proposal mechanism, this is for more advanced books) gives the results in Figure 7.11 which show results similar to the rejection sampler except that is takes many fewer iterations, namely approximately 8000, to generate 150 Earth models (or about 50 flow simulations per posterior Earth model).

7.4.4 Optimization Methods

Not all techniques for finding inverse solutions use Bayes' rule as a guide. These techniques are generally termed optimization methods, they are often much more efficient than the sampling techniques, such as rejection sampler and Metropolis sampler, but generally do not produce realistic models of uncertainty. In fact, the aim of optimization techniques is not to model uncertainty, it is simply to find an Earth model that matches the data and has some other desirable properties. Optimization techniques such as gradient-based optimization, Ensemble Kalman filters, genetic algorithm, simulated annealing or search methods such as the neighborhood algorithm focus directly on the data matching problem:

$$\text{Find } \mathbf{m} \text{ such that } g(\mathbf{m}) = \mathbf{d}$$

or more generally

$$\text{Find } \mathbf{m} \text{ such that } P(\mathbf{D} = \mathbf{d} | \mathbf{M} = \mathbf{m}) \text{ is maximal}$$

or in the context of a Gaussian type likelihood:

$$\text{Find } \mathbf{m} \text{ such that RMSE}(\mathbf{m}) \text{ is minimal}$$

The basic difference between a sampling technique such as rejection sampling and an optimization method lies in the way $\mathbf{m}$ is changed and how such change is accepted or rejected. Recall that during sampling $\mathbf{m}$ is changed consistent with any prior information. In gradient-based optimization methods for example, the change is made such that a decrease of RMSE (or increase of likelihood) is obtained for each iteration step (techniques on how to do this are not discussed here; this is for optimization books to elaborate on). However, the decrease determined by the optimization method (i) depends on the particular method of optimization used (e.g., what gradient method is used) and (ii) does not necessarily account for prior information (such as for example expressed in a training image). In other words, one may end up generating an inverse solution **m** that is not even part of the initial prior set, hence violating Bayes' rule as explained above. It should also be realized that simply obtaining many solutions to an inverse problem does not entail that one has a realistic model of uncertainty. One could, for example, apply the same optimization algorithm with different initial starting models, then observe that multiple different solutions are obtained from these different starting points. This observation only shows that there are many solutions, however these solutions do not necessarily (actually rarely) represent a realistic model of uncertainty under the auspices of Bayes' rule.

In this book, on modeling uncertainty, prior information is considered highly valuable, particularly if there is large uncertainty on parameters that are sensitive to a decision problem. If there is only small uncertainty, then models of uncertainty may not be relevant to solving the problem in the first place. In that case, optimization methods are preferred.

Further Reading

Dubrule, O. (2003) *Geostatistics for seismic data integration in Earth models.* SEG 2003 Distinguished Instructors Short Course.

Strebelle, S., Payrazyan, K., and Caers, J. (2003) Modeling of a deepwater turbidite reservoir conditional to seismic data using multiple-point geostatistics. *SPE Journal*, **8**(3), 227–235.

Tarantola, A. (2005) *Inverse Problem Theory and Method for Model Parameter Estimation*, SIAM publications.

8

Modeling Structural Uncertainty

Co-authored by Guillaume Caumon

Nancy School of Geology, France

Uncertainty modeling being the main topic in this book brings us to the evident question: how to build multiple structural models that serve as a representation or model of uncertainty? This is not as trivial a question as the modeling of properties on a regular grid. The various constraints of geological consistency as well as the difficulty of automating the construction of structural models make this a difficult task.

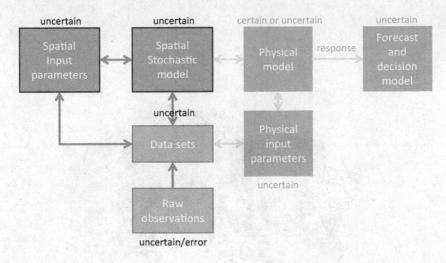

8.1 Introduction

Previous chapters dealt with the spatial modeling of properties/variables that can be represented on a simple Cartesian grid. Various models of spatial continuity that are used

---

Modeling Uncertainty in the Earth Sciences, First Edition. Jef Caers.
© 2011 John Wiley & Sons, Ltd. Published 2011 by John Wiley & Sons, Ltd.

in generating several alternative Earth Models, potentially constrained to data, were presented. What was ignored is the structural framework and grid needed to simulate these properties on. In some cases, the framework is a simple grid, surface or volume, for example a grid representing the layers in the Earth's atmosphere. However, in many cases, particularly in modeling the subsurface, the structure modeled can be quite complex (see Figure 8.1) and requires subdividing the domain of interest into units based on geological considerations (setup process, age, rock type). These units are bounded by 3D surfaces such as faults and horizons, which can be modeled from the observations at hand, then serve as a structural framework to create a conforming grid. In addition, one may want to model the change of these surfaces over geological history, hence a change in the grid over time.

The representation of the structural framework appeals to classical CAD techniques such as parametric lines and surfaces and/or meshes. In addition, modeling of geological surfaces relies on specific methods to honor typical subsurface data types and guarantee the consistency of the model with geological processes.

When it comes to modeling geological surfaces and considering the related uncertainty, it is important to distinguish between the three following characteristics:

1 The topology, which describes the type of surface (spherical, donut-shaped, open, with hole(s), etc.), and the connectivity between surfaces, for instance the termination of a fault onto another fault. The topology of an object does not change (it is invariant) when the object undergoes a continuous deformation (Figure 8.2).

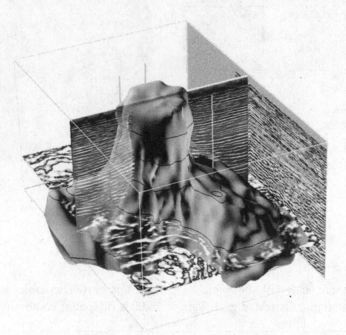

Figure 8.1 Complex structure (salt dome) interpreted from seismic data.

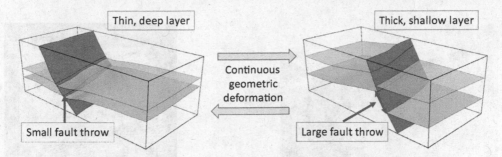

Figure 8.2 Two geological surface models having different geometry but the same topological features (number of surfaces and adjacency between those).

2 The geometry, which specifies the 3D position and shape of the surface in space. In a numerical earth model, it is generally given by the location of some nodes and by an interpolation method between nodes (linear, spline, etc.).

3 The properties, or attributes, attached to the object. These can be rock properties (porosity, soil type, etc.), physical variables (temperature and pressure) or geometrical properties (e.g., local slope or curvature).

This chapter concentrates on building structural models for the subsurface (local or global scale) and dealing with uncertainty related to geological structures. Because uncertainty about properties has been considered in previous chapters, the focus is mostly on uncertainty related to the geometry and to the topology of structural models. So this chapter is quite specific to the data and problems related to modeling the subsurface; however, other Earth Science applications could be envisioned with these techniques, for example classification of fossils or mapping the Earth surface from remote sensing data. The goal of this chapter is not to provide a detailed perspective on building structural models, but to introduce and categorize those elements of structural modeling that are most subject to uncertainty, hence play a key role in modeling uncertainty in 3D Earth models.

8.2 Data for Structural Modeling in the Subsurface

Data most used for structural modeling are geophysical images obtained through geophysical surveys such as seismic surveys (Figure 8.3) or EM (electromagnetic surveys). These can be ground based or airborne, even using satellites (synthetic aperture radar data) for example to detect ground movement.

Geophysical data provide a complete coverage of the subsurface along one section (e.g., 2D seismic) or over a whole volume (e.g., 3D seismic). The geophysical data used are the outcome of a complex chain of geophysical processing. For example, seismic acquisition is based on the emission of an artificial vibration (the source) either onshore or offshore (Figure 8.3), whose echoes are recorded by a set of geophones. The seismic waves emitted by the source undergo refraction and reflection when propagating through

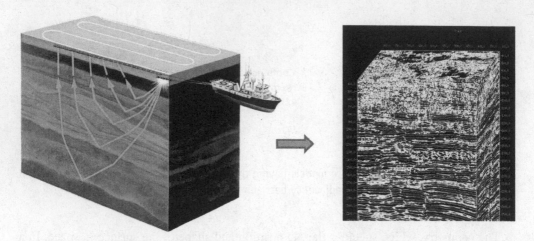

Figure 8.3   Configuration of sources and receivers in a marine seismic survey (left). The resulting 3D processed seismic data cube on which interpretations are made (right).

rocks of different nature. This signal carries the contrasts of impedance (the product of seismic wave velocity by the rock density) as a function of time (the travel time between the source and the geophone). Seismic processing turns the raw data into a useable 3D seismic image, as shown in Figure 8.3.

The processing of seismic data is very complex and computationally very demanding, requiring a number of corrections and filtering operations, whose parameters are generally inferred from the raw seismic data and calibrated at wells. For example, the conversion of travel times to depth is ambiguous, since it requires an estimation of seismic velocities which are not known a priori. Therefore, seismic data are not precise, especially in the vertical direction. Also, they have a poor resolution (from a few meters for shallow, high resolution seismic to about 20–50 meters for classical surveys). Lastly, the significance of seismic images decreases as depth increases, due to the attenuation of the signal amplitude. Notwithstanding these limitations, the value of seismic data is its ability to cover an entire volume. The fuzzy picture provided by seismic is of paramount importance for structural modeling, as it provides a view of 3D geometries. Seismic amplitudes are routinely used to extract significant surfaces such as horizons, unconformities, faults, and so on (Figure 8.4). As the seismic image is fuzzy, such extraction requires tedious interpretation through manual picking. In other words, an interpreter sits for hours in front of a computer screen using software to pick points that he/she thinks correspond to geologic features. These points will then help create surfaces and eventually a structural framework.

8.3 Modeling a Geological Surface

The typical input for structural modeling is a set of samples in 3D that represents the most likely locations where the surface exists. Before considering how uncertainty can be

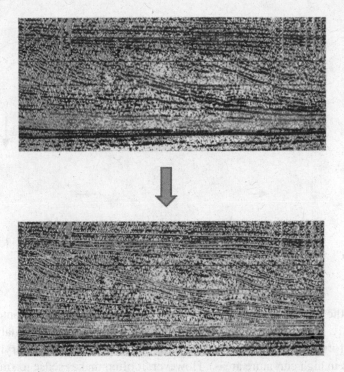

Figure 8.4 A 2D slice from a 3D seismic data cube and its interpretation by an expert.

addressed, we will first provide some details about how to represent a possible structural model constrained to these data.

A structural model is an idealization of the main rock boundaries, generally represented by a discrete set for each surface. For example, a continuous surface is often approximated by a set of planar polygons forming a 2D mesh in 3D space. The creation of one surface from interpretation points goes through several steps, summarized by Figure 8.5:

A Establish the point set to work with.

B-C Determine the lateral extent of the surface, for instance take the convex hull of the data points (imagine an elastic band stretching around the data points), or simply take the intersection of the average plane with the domain of study.

D Create a first surface model from the points outline and/or the average plane, with a specific level of detail

E-F Deform it to make it closer to the data.

Although these steps are often automated by software, there are two main modeling decisions to be made during this workflow which are prone to uncertainty. Firstly, the

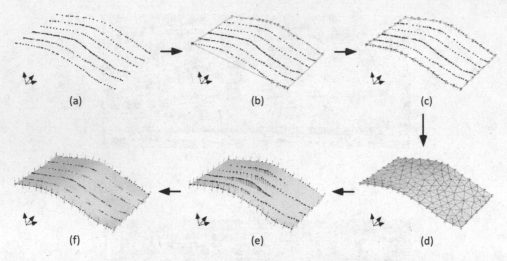

Figure 8.5 Fitting a surface to a set of points.

resolution of the surface tells how much geometric detail may be represented in the final surface. Ideally, a model should have a minimal size to reflect the available data without loss of information (e.g., few triangles and points in at smoothly varying areas, and high densities in high curvature areas). However, it often makes sense to smooth out high resolution details considered as noise by choosing a surface resolution coarser than the interval between data. Secondly, away from the points, the surface is generally set to be smooth, to reflect the principle of minimum energy (or second law of thermodynamics) during rock deformation and rupture when the rocks are buried in the subsurface. The balance between smoothness and data fitting can be controlled by the modeler, conveying how much one trusts interpretation points versus simplified physical principles. The resolution and degree of smoothness of the resulting surface are both contributing to the uncertainty about the structural model.

8.4 Constructing a Structural Model

Discontinuities within the subsurface are due to changes of depositional conditions, erosion, tectonic events such as faulting and folding or late transport of reactive fluids. Previously, ways to create 3D surfaces to represent the 3D geometry of discontinuities, for example, horizons and faults were described. In constructing a structural model, we can no longer deal with surfaces one at a time, but need to consider how they are related to each other. Modeling correctly the relationships between geological surfaces is critical to ensure the consistency of a numerical model as formulated by geological validity conditions and a set of geometrical and topological constraints to enforce these conditions.

8.4.1 Geological Constraints and Consistency

Numerical modeling of 3D objects can produce shapes which are mathematically correct but do not represent any valid natural object (Figure 8.6). Since Earth data are often sparse

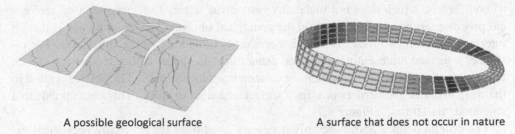

A possible geological surface   A surface that does not occur in nature

Figure 8.6   An orientable and non-orientable surface (Moebius ring).

and noisy, 3D modeling relies on automatic and interactive consistency checks such that valid structural models are generated.

Previously, the creation of surfaces constrained to point sets in three dimensions was presented. These data points often mark the common boundary of two volumes of different nature, for example a contrast of a certain property or the existence of a certain discontinuity. As such, a geological interface can be seen as a magnet whose sides have a different polarization. Such a surface with two well defined sides is said to be orientable. A famous example of non-orientable surface is the Moebius ribbon depicted in Figure 8.6; it cannot be created by a natural process.

Each side of a surface bounds a particular layer or (fault) block. For volume consistency, a surface must not self-intersect. More generally, the volumes defined by a set of interfaces must not overlap. For instance, the hatched part in Figure 8.7 would belong

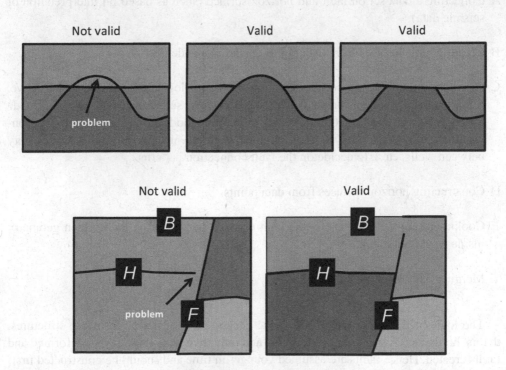

Figure 8.7   Valid and non-valid configuration of faults and horizons in the subsurface.

to both layers, which does not make any geological sense. The succession of geological processes imposes a hierarchy on the geological objects. One consequence is that an interface between two rock volumes is necessarily bounded by other interfaces. For instance, a surface representing a horizon cannot float in space: its borders have to lie on other natural surfaces, such as a fault or domain boundary surface. The only exception to that rule is faults: the borders of a fault surface can float in space. This corresponds to a so-called "null throw" (Figure 8.7).

The criteria defined above are only necessary conditions for enforcing geological validity. Defining sufficient conditions is much more difficult and calls for higher level concepts in structural geology and sedimentology. Enforcing such conditions intrinsically in the model building process is neither always easy nor applicable, and requires a careful prior geological analysis of the domain being studied. Alternatively, models can be checked for consistency with some type of data or physical principles. This a posteriori check is usually computationally expensive.

8.4.2 Building the Structural Model

Given our knowledge about construction of a surface and the various conditions required to build a consistent structural model, the following basic steps can be used to build such model (Figure 8.8):

A Collecting a data set on fault and horizon surface (such as based on interpretation of seismic data).

B Creating fault surfaces from data points, each independently.

C Connecting the faults properly accounting for the geological consistency rules above. Significant uncertainty is present at this stage, because subsurface data seldom provide a clear image close to discontinuities. It is up to the modeler, based on analog reasoning and careful data analysis (e.g., regional tectonic history, hydraulic connectivity between wells, etc.), to decide on the fault connection patterns.

D Constructing horizon surfaces from data points.

E Cookie-cutting the horizons by the fault network and updating the horizon geometry inside each fault block.

F Merging fault and horizon surfaces.

The logic behind these steps follows the genetic rules of most subsurface structures, that is, horizons (layers) were created first, and only then were these layers deformed and faults created. Hence faults are assumed younger in time and should be constructed first.

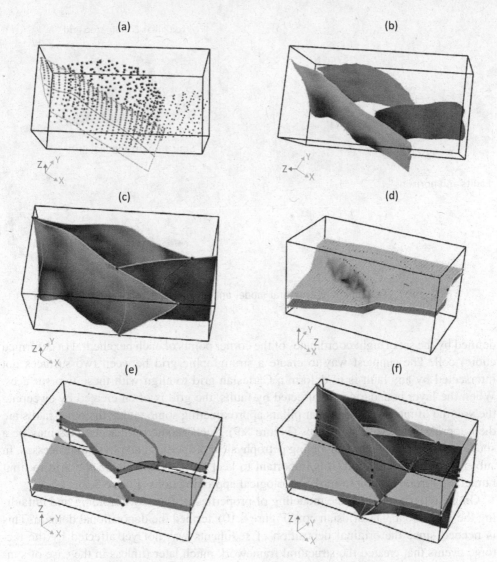

Figure 8.8 Sequence of steps in structural modeling.

8.5 Gridding the Structural Model

8.5.1 Stratigraphic Grids

Most Earth models consist of both a structural framework or model and a set of properties existing within this framework. To assign properties and solve partial differential equations that govern physical processes in the subsurface, a grid needs to be constructed in this structural model.

For example, stratigraphic grids (Figure 8.9) are like 3D Cartesian grids except that they are deformed to fit the stratigraphy (layering) of the subsurface. They are uniquely

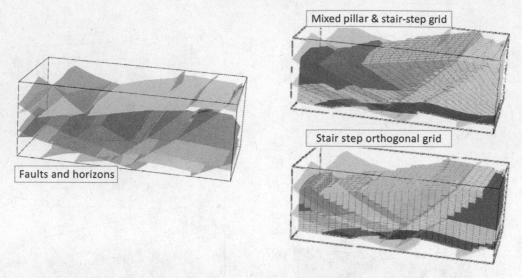

Figure 8.9 Example of a structural model and two possible stratigraphic grids.

defined by the set of eight coordinates of the corner points of each hexahedral (a deformed cubic) cell. The simplest way to create a stratigraphic grid between two surfaces not intersected by any fault is to deform a Cartesian grid to align with these two surfaces. When the layer boundaries are affected by faults, the grid is often created by sweeping the volume of interest along linear pillars approximating some faults; the other faults are then approximated with stair steps (Figure 8.9). Such geometric inaccuracies may be a source of uncertainty when modeling petrophysical properties and physical processes. In subsurface flow simulation, it is important to keep orthogonal cells in the grid to limit numerical errors, leading to further geological approximations (Figure 8.9).

On a stratigraphic grid, the modeling of properties is done by unfolding and unfaulting the grid into a 3D Cartesian box (Figure 8.10), termed the depositional domain. This is needed since the original deposition of sediments was not yet affected by the tectonic events that created the structural framework much later (unless in the case of synsedimentary tectonic events). Once this mapping into a depositional environment has been done, spatial modeling such as discussed in Chapters 5–7 can take place. Note that this also means that any data need to be mapped into this depositional domain. After properties are modeled (Figure 8.8) they are mapped back to the physical domain.

In magmatic and metamorphic rocks, such mapping is often impossible, so the structural framework is used to identify statistically homogeneous regions and possibly to define local directions of anisotropy for petrophysical modeling.

8.5.2 Grid Resolution

In creating a grid, it is necessary to decide on the grid resolution, that is, what the average size of grid cells in the Earth model is. A high resolution grid may be able to represent the

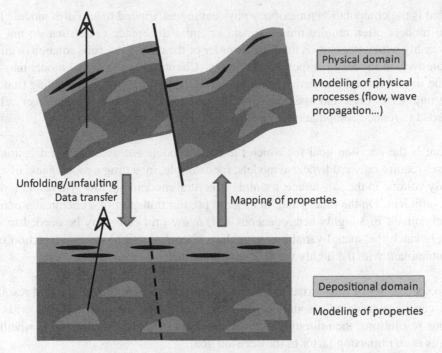

Physical domain

Modeling of physical processes (flow, wave propagation...)

Unfolding/unfaulting
Data transfer

Mapping of properties

Depositional domain

Modeling of properties

Figure 8.10 Assigning properties in a stratigraphic grid by creating a "depositional domain."

modeled phenomenon more accurately, but may be too cumbersome computation-wise. Several factors should be considered in deciding on the grid resolution:

- What are the smallest scale data available? Soil samples, cores, well-log measurements are often direct measurements of the studied phenomenon. However, given the volume/size of such data, even a small Earth model requires several billion grid cells to allow direct inclusion into the model grid. A grid cell size considerably larger than the dimension of the smallest scale data often needs to be considered. This means that finer scale variability is ignored or represented as an equivalent second order tensor property. In that case, either small scale data are assumed to be representative of the entire grid cell in which it is located, or some averaging methodology needs to be performed (requiring a change of scale of four to five orders of magnitude typically). This loss of information should be considered as a source of uncertainty.

- What is the smallest scale phenomenon that should be represented? Small scale barriers (inches thick) are important to flow in the subsurface, or contact between saline and fresh continental waters may be critical to ocean modeling. Ideally, such phenomena should be represented in an Earth model at their respective scales. However, this is not always feasible because their size is too small for current computational capabilities. Similarly, some form of averaging or implicit representation is needed (see also Chapter 3), leading to an additional source of uncertainty.

- What is the computation time of any physical model applied to an Earth model? Physical models often require finite element or finite difference codes that do not scale favorably with computation time (in the order of the number of cells squared or cubed). Moreover, in modeling response uncertainty (Chapter 10) a physical model may need to be run on several alternative Earth models. Often some form of upscaling (reducing the number of cells and assigning upscaled/averaged properties to the larger cells) is needed to reduce computation time to realistic levels.

- What is the decision goal for which the Earth models are used? Some decision purposes require only crude/coarse models; for example, in getting a rough guess of an ore body volume in the subsurface, a simple structural model with low grid resolution may be sufficient. On the other hand, in order to predict transport of a chemically complex contaminant in a highly heterogeneous soil, many grid cells may be needed to accurately catch the spatial variability as well as chemical and physical interaction of the contaminant with the highly variable environment.

No specific set of rules exists to choose the grid resolution. In fact, the grid resolution itself can be considered a source of uncertainty and one may opt to build grids with various resolutions, then through a sensitivity study (Chapter 10) determine whether or not this is an impacting factor to the decision goal.

8.6 Modeling Surfaces through Thicknesses

In the previous sections, a fairly general approach to modeling structures was discussed. This approach has been traditionally applied to subsurface structures but could be applied to any surface modeling application requiring a representation of both geometry and topology. In certain applications, such as modeling specific depositional environments in the subsurface, more specialized forms of modeling can be used that are tailored to the application and generally less complex. This is the case when modeling sedimentary structures that have not undergone any deformation, or for which the deformation part has been accounted for by unfolding and unfaulting the structures. Sediments are deposited on top of each other, that is, in a sequence of depositional layers over a depositional plane with periods of erosion that have removed material. A simple way to represent a surface is to start from a 2D thickness map (Figure 8.11). This thickness variable is basically a 2D variable that can simply be represented on a Cartesian grid. Surfaces created by depositional events can easily be represented in this fashion. Stacking various thickness outcomes on top of each other results in a 3D volume, as shown in Figure 8.11. Erosion is simply a negative thickness. In this way, surfaces can be modeled using the Cartesian grid-based techniques described in Chapters 5–7.

8.7 Modeling Structural Uncertainty

So far we have concentrated on building a single structural model. Often such models require a large amount of manual intervention to make them consistent with the ideas the

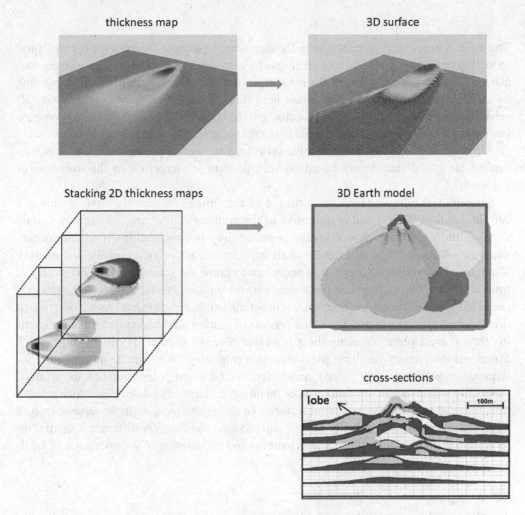

Figure 8.11 Turning thickness maps into volumes bounded by surfaces that stack on top of each other. These figures show how lobes of sand stack on top of each other in a shale background.

modeler has in mind. Uncertainty modeling being the main topic in this book brings us to the evident question: how to build multiple structural models that serve as a representation or model of uncertainty? This is not as trivial a question as the modeling of properties on a regular grid. The various constraints of geological consistency as well as the difficulty of automating the construction of structural models make this a difficult task. So far, software exists to perturb the geometry of structural models around some reference, but no commercial software is available in constructing hundreds of structural models with varying topologies: only research ideas and implementations exist. Nevertheless, in this section firstly the sources of uncertainty related to such modeling are outlined, and then, in the next section, how such models of uncertainty are generated practically is briefly mentioned.

8.7.1 Sources of Uncertainty

The main sources of uncertainty lie in the data source (seismic) being used for structural modeling and the interpretation of such data by a modeler. A short overview of these two aspects is provided. The order of the magnitude of uncertainty can be different depending on the data acquisition condition in the field (i.e., land data or marine data, 2D or 3D seismic, etc.), the subsurface heterogeneity, and the complexity of the structural geometry. For example, land data generally provide poorer seismic data than marine data.

Although it is difficult to set a universal rule, a typical example of hierarchy in subsurface structural uncertainty based on seismic data is suggested in the overview of Figure 8.12.

Seismic data basically measure a change of amplitude of a seismic wave in time at a certain position. This signal is indicative of the contrast of rock impedance at a certain depth in the subsurface. A first source of uncertainty in structural interpretation occurs when two different rock units have a small impedance contrast (e.g., granite and gneiss). Then, to go from time to depth, it is necessary to know the seismic velocity (velocity × time = depth), that is, how fast these waves travel through the subsurface. This velocity needs to be somehow determined, since it is not measured directly; hence such determination is subject to uncertainty. The entire process of putting all the recorded seismic signals in "their correct place" (moving them from one place to another) is called "migration". Structural uncertainty resulting from uncertain migration (the signals on which the interpretation was done were placed incorrectly) can be a first order structural uncertainty, especially when seismic data are of poor quality (i.e., large uncertainty in velocity determination due to a heterogeneous subsurface). In such situations, multiple seismic images migrated using different velocity models can produce significantly different structural interpretations that exhibit different fault patterns and appearance or disappearance of faults

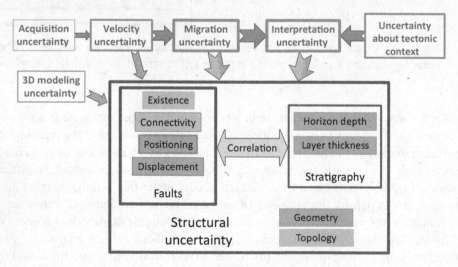

Figure 8.12 Main factors and types of structural uncertainty, with typical significance figured as frame and arrow width.

Migration
Uncertainty

Seismic image 1   Seismic image 2

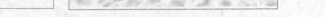

Figure 8.13   Migration uncertainty leading to several alternative seismic data sets.

depending on which seismic image is considered. Hence, the image on which modelers perform their interpretations is uncertain to start with (Figure 8.13).

With poor seismic data, a single seismic image can produce considerably different structural interpretations depending on the different decisions made on horizon/fault identification. Such an uncertainty from structural interpretation can be first order, especially when the structural geometry is complex, since multiple interpretations can exhibit a significant difference in amount of faults or fault pattern (human bias in interpretation). This uncertainty is modeled by providing multiple possible alternatives of structural interpretations (Figure 8.14). These may be defined within one or several tectonic scenarios (a succession of tectonic events through geological time and hypothesis about rock deformation mechanisms).

Correlating horizons across a fault can be difficult unless wells are available on both sides of the fault, since a "wrong" pair of reflectors can be picked as indicating the same

Interpretation
Uncertainty:
fault network

Interpretation 1 on seismic image 1   Interpretation 2 on seismic image 1

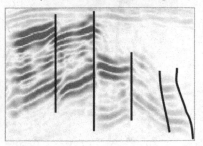

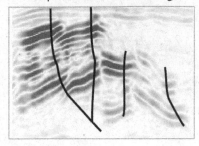

Figure 8.14   Multiple fault interpretations based on one seismic image.

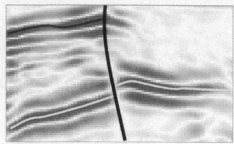

Figure 8.15   The blue "reflectors" can be correlated differently across the fault, as indicated by the various colored lines. This leads to a different interpretation in fault displacement.

horizon (Figure 8.15). Erroneous horizon identification would lead to misinterpreting the fault throw (vertical component of the fault slip).

The uncertainty in horizon position (Figure 8.16) is attributed to (i) the error in selecting a reflector due to the low resolution of the seismic data, and (ii) uncertainty in time–depth conversion. Because layer thickness uncertainty is generally smaller than depth uncertainty, it is often suitable to sample the uncertainty of a set of horizons at once rather than the uncertainty of each horizon separately.

The magnitude of uncertainty related to fault positioning depends on the quality of the seismic image on which the interpretation is done, thus its magnitude is evaluated through visual inspection of the seismic image (Figure 8.16). This uncertainty is also of lower order importance compared to the uncertainty related to the fault identification or the uncertainty in fault connectivity. It usually increases with depth, and may be locally smaller if the fault has been identified along boreholes (Figure 8.17).

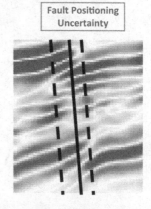

Figure 8.16   Uncertainty in the position of horizon and fault position.

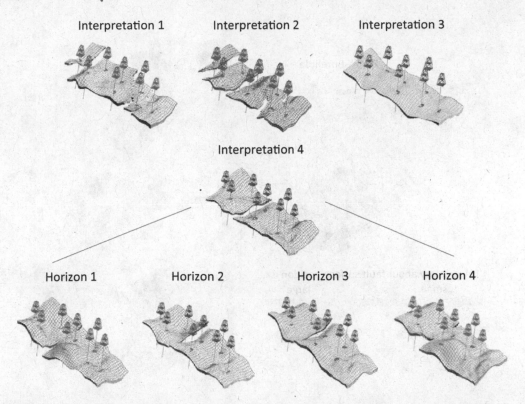

Figure 8.17 Eight structural models from a set of 400 structural models representing structural uncertainty in an oil reservoir case study. The four bottom most models have the same interpretation 4, but each exhibit a different top horizon.

8.7.2 Models of Structural Uncertainty

Models of structural uncertainty are like most other models of uncertainty in this book: a set of alternative structural models is generated based on the various sources of uncertainties identified (Figure 8.17). Some sources of uncertainty are discrete in nature, such as for example the seismic image used (a few may be chosen) or the structural interpretation (a few fault scenarios may be chosen), while others are more continuous in nature, for example, the position of a particular horizon or fault may be modeled by specifying an interval around the surface within which the position is varied in a specific way (Figure 8.18). For example, a simple way to stochastically vary a horizon surface is to vary the thickness map that represents this surface.

However, various practical issues arise in building such models of structural uncertainty that prevent wide availability on software platforms. One of the main reasons lies in the difficult automation for generating structural models, particularly when the models are structurally complex. Therefore, most existing tools operate by perturbing directly a stratigraphic grid rather than directly perturbing or simulating a structural model, thereby possibly reducing uncertainty.

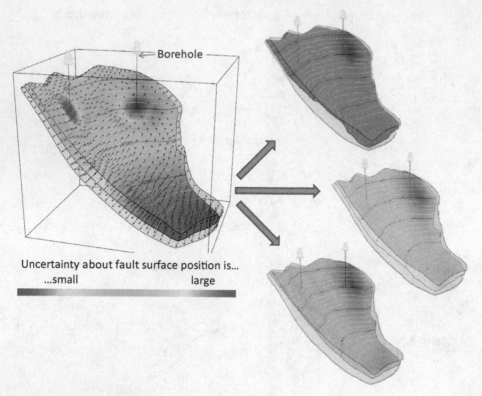

Figure 8.18 Uncertainty envelope about a fault surface and three possible geometries for that surface.

Sampling topological uncertainties is still at the research level, because it demands to replace all the manual editing performed in structural model building by ancillary expert information. For example, stochastic fault networks can be generated from statistical information about fault orientation and shape, relationships between fault size and fault displacement, and truncation rules between fault families (Figure 8.19).

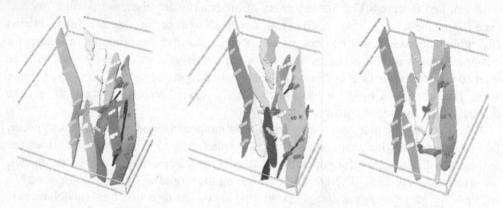

Figure 8.19 Three possible fault networks constrained to 2D seismic interpretations and exhibiting the same prior statistics of orientation and size/displacement ratio for fault families.

Further Reading

Bond, C.E., Gibbs, A.D., Shipton, Z.K., and Jones, S. (2007) What do you think this is? "Conceptual Uncertainty" in geoscience interpretation. *GSA Today*, **17**, 4–10.

Caumon, G. (2010) Towards stochastic time-varying geological modeling. *Mathematical Geosciences*, **42**(5), 555–569.

Cherpeau, N., Caumon, G., and Levy, B. (2010). Stochastic simulation of fault networks from 2D seismic lines. *SEG Expanded Abstracts*, **29**, 2366–2370.

Holden, L., Mostad, P., Nielsen, B.F., *et al*. (2003) Stochastic structural modeling. *Mathematical Geology*, **35**(8), 899–914.

Suzuki, S., Caumon, G., and Caers, J. (2008) Dynamic data integration into structural modeling: model screening approach using a distance-based model parameterization. *Computational Geosciences*, **12**, 105–119.

Thore, P., Shtuka, A., Lecour, M., *et al*. (2002) Structural uncertainties: Determination, management, and applications. *Geophysics*, **67**, 840–852.

9
Visualizing Uncertainty

We cannot process hundreds of models in climate modeling when we know one such model may take several days of computing time. We cannot run flow simulation on thousands of reservoir models to evaluate whether drilling a new well is worth the cost. The good news is that we don't have to. We can be selective of the models that we use to process. However, this requires that we get a better visual insight in the uncertainty represented by those thousands of models and in this chapter some important tools to do so are discussed.

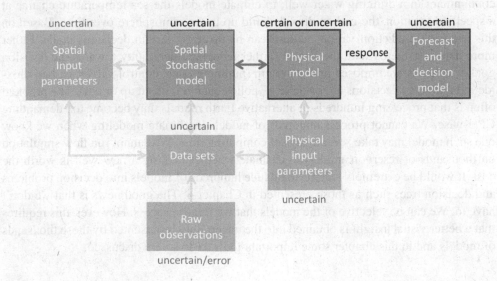

9.1 Introduction

The purpose of Earth modeling and prediction is very clear: produce forecasts (climate, reservoir flow, contaminant transport, groundwater recharge efficiency), estimate reserves (mineral resources, total contaminated sediment volume) or make decisions (choose where to drill well, decide whether to obtain more data, decide to clean up, change

Modeling Uncertainty in the Earth Sciences, First Edition. Jef Caers.
© 2011 John Wiley & Sons, Ltd. Published 2011 by John Wiley & Sons, Ltd.

policies). As observed so far, many fields of expertise in modeling are required to complete this task: spatial modeling, structural modeling, process modeling, geological interpretation, data processing and interpretation, modeling of partial differential equations, inverse modeling, decision theory, and so on. In many applications, a rigorous assessment of uncertainty is critical since it is fundamental to any geo-engineering decision making process.

In this book several techniques to model uncertainty about the Earth's properties and structures have been discussed. We saw how we can represent such uncertainty by generating many alternative Earth models, possibly hundreds or thousands. Now what do we do with them? How do we start incorporating them into a decision making process? How do we process them for prediction purposes? Regarding the last question, the term "applying a transfer function" has been coined for taking a single model and calculating a desired or "target" response from it. If many Earth models have been created, the transfer function needs to be applied to each Earth model, resulting in a set of alternative responses that model/reflect the uncertainty of the target response. This is a common Monte Carlo approach. In a reservoir/aquifer context, this response may be the amount of water or oil from a new well location; in mining this could be a mining plan obtained through running an optimization code; in environmental applications this may be the amount of contaminant in a drinking water well; in climate models the sea temperature change at a specified location, the carbon dioxide build up in the atmosphere over time. Based on this response prediction various actions can be taken or certain decisions made. Either more data is gathered with the aim of further reducing uncertainty towards the decision goal or controls are imposed on the system (pumping rates, control valves, carbon dioxide reduction) or decisions are made (e.g., policy changes, clean up or not). The problem often is that processing hundreds of alternative Earth models may become too demanding CPU-wise. We cannot process hundreds of models in climate modeling when we know one such model may take several days of computing time. We cannot run flow simulation on thousands of reservoir models to evaluate whether drilling a new well is worth the cost. It would be extremely tedious to include hundreds of models into decision problems and decision trees such as those presented in Chapter 4. The good news is that we don't have to. We can be selective of the models that we use to process. However, this requires that a better visual insight is obtained into the uncertainty represented by those thousands of models and in this chapter some important tools to do so are discussed.

9.2 The Concept of Distance

An observation critically important to understanding the uncertainty puzzle is that the complexity and dimensionality (number of variables) of the input data and models are far greater than the complexity and dimensionality of the desired target response. In fact, the desired output response can be as simple as a binary decision question: do we drill or not, do we clean up or not? At the same time the complexity of the input model can be enormous, containing complex relationships between different types of variables, physics (e.g., flow in porous media or wave equations) and is, moreover, varying in a spatially complex way. For example, if three variables (soil type, permeability and porosity) are simulated on a one million cell grid and what is needed is contaminant concentration at 10 years for a groundwater well located at coordinate (x,y), then a single input model has dimension of 3×10^6 while the target response is a single variable.

This simple but key observation suggests that the uncertainty represented by a set of Earth models can be represented in a simpler way, if the purpose of these Earth models is taken into account. Indeed, many factors may affect contaminant concentration at 10 years to a varying degree of importance. If a difference in value of a single input variable (porosity at location (x,y,z)) leads to a considerable difference in the target response, then that variable is critical to the decision making process. Note that the previous statement contains the notion of a "distance" (a difference of sorts). However, because Earth models are of large dimension, complex and spatially/time varying, it may not be trivial to discern variables that are critical to the decision making process easily. To make the uncertainty puzzle simpler, the concept of a distance is introduced. A distance is a single, evidently positive value that quantifies the difference between any two "objects." In our case the objects are two Earth models. If there are L Earth models, then a table of L × L distances can be specified. The mathematical literature offers many distances to choose from; a very common distance that is introduce later is the Euclidean distance (in 2D it would be a measure of the distance between two geographic locations on a flat plane). The choice between distances provides an opportunity to choose a distance that is related to the response differences between models. This will allow "structuring" uncertainty with a particular response in mind and create better insight into what uncertainty affects the response most. Earth models can be considered as puzzle pieces: if two puzzle pieces are deemed similar then they can be grouped and represented by some average puzzle piece (the sky, the grass, etc.). This, however, requires a definition of similarity. This is where the distance comes in. In making that distance a function of the desired response, the grouping becomes effective for the decision problem or response uncertainty question we are trying to address. For example, if contaminant transport from a source to a specific location is the target, then a distance measuring the connectivity difference (from source to well) between any two Earth models would be a suitable distance.

Defined in the next section are some basic concepts related to distances that allow the uncertainty represented by a large variety of models to be analyzed very rapidly. As an initial illustration of how a distance renders complex phenomena more simple, consider the example in Figure 9.1. As discussed in Chapter 8, uncertainty in structural geometry is complex and attributed to various sources. Figure 9.2 shows a few structural models from a case study whereby a total of 400 structural Earth models are built. As discussed in Chapter 8, a structural model consists of horizon surfaces cut by faults. To distinguish between any two structural models, the joint difference between corresponding surfaces in a structural model is then used as a distance, termed d_{H}. Figure 9.1 explains how this is done exactly. A surface consists of points x,y with a certain depth z (at least for surfaces of non-overhang structures). The joint distance between these depth values z for each surface of a model k and the same surface of a model ℓ is then a measure of the difference in structural model. How this distance is actually calculated is not the point of discussion for this book. Figure 9.1 shows that such difference depends on the difference in fault structures as well as the difference in horizons (as was outlined in Chapter 8).

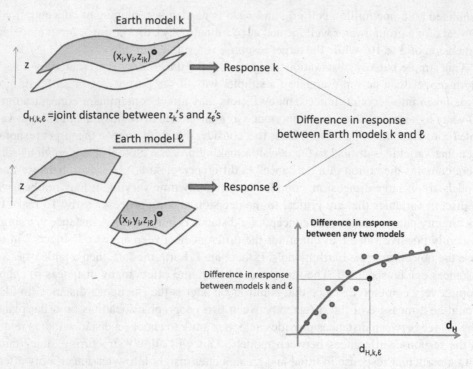

Figure 9.1 Calculating the difference between two structural models and comparing it with the difference in response calculated from such models.

To evaluate whether this distance can actually provide a better insight into the difference in cumulative oil production response between any two models, the squared difference in response is also calculated and this difference plotted versus d_H in Figure 9.1. If this is done for all pairs of models and the moving average taken, the smooth line in Figure 9.1 is obtained. Note that the latter plot is nothing more than the variogram of production response, as defined in Chapter 5; however, the distance on the x-axis of this variogram plot is the geometrical distance between any two models. Recall that the variogram is a measure of dissimilarity, so small variogram values mean that samples close by (in the distance defined) are related to each other. If the distance was not informative about a difference in production response, then this variogram would be a pure nugget variogram. Clearly this is not the case, as shown in Figure 9.2, for the actual case. The distance is informative about production response difference and provides a way to structure variability between a set of complex structural models in a simple fashion.

9.3 Visualizing Uncertainty

9.3.1 Distances, Metric Space and Multidimensional Scaling

The notion of distance can now be put in a more mathematical context. The equations are given here for completion; what is important, however, are the plots that result from them.

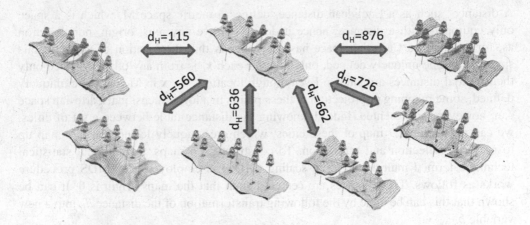

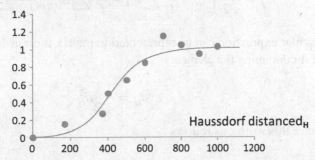

Figure 9.2 Seven structural Earth models out of 400 are shown (top). The distance d_H between a single model and the other six is plotted; (bottom) the variogram of production response as function of the distance d_H.

A single (input) Earth model i is represented by a vector $\mathbf{x}_i$, which contains either the list of properties (continuous or categorical or a mix of them) in each grid cell or an exhaustive list of variables uniquely quantifying that model. The "size" or "dimension" N of the model is then the length of this vector, for example, the number of grid cells in a gridded model. N is typically very large. L denotes the number of alternative models generated with typically $L \ll N$. All models are collected in the matrix X:

$$X = [\mathbf{x}_1 \, \mathbf{x}_2 \ldots \mathbf{x}_L]^T \text{ of size } L \times N \qquad (9.1)$$

One of the most studied distances is the Euclidean distance, which is defined as

$$d_{ij} = \sqrt{(\mathbf{x}_i - \mathbf{x}_j)^T (\mathbf{x}_i - \mathbf{x}_j)} \qquad (9.2)$$

if this distance were applied to a pair of models $\mathbf{x}_i, \mathbf{x}_j$. Mathematically, one states that models exist within a Cartesian space D (a space defined by a Cartesian axis system) of high dimension N, each axis in the Cartesian grid representing the value of a grid cell.

A distance, such as a Euclidean distance, defines a metric space M, which is a space only equipped with a distance; hence it does not have any axis, origin, nor direction as, for example, a Cartesian space has. This means that the location of any $\mathbf{x}$ in this space cannot be uniquely defined, only how far each $\mathbf{x}_i$ is from any other $\mathbf{x}_j$, since only their mutual distances are known. Even though locations for $\mathbf{x}$ in M cannot be uniquely defined, some mapping or projection of these points in a low dimensional Cartesian space can, however, be presented. Indeed, knowing the distance table between a set of cities, we can produce a 2D map of these cities, which are uniquely located on this map up to rotation, reflection and translation. To construct such maps, a traditional statistical technique termed multidimensional scaling (MDS) is employed. The MDS procedure works as follows. The distances are centered such that the maps origin is $\mathbf{0}$. It can be shown that this can be done by the following transformation of the distance d_{ij} into a new variable b_{ij},

$$b_{ij} = -\frac{1}{2}\left(d_{ij}^2 - \frac{1}{L}\sum_{k=1}^{L}d_{ik}^2 - \frac{1}{L}\sum_{l=1}^{L}d_{lj}^2 - \frac{1}{L^2}\sum_{k=1}^{L}\sum_{l=1}^{L}d_{kl}^2\right)$$

This scalar expression can be represented in matrix form as follows. Firstly, construct a matrix A containing the elements

$$a_{ij} = -\frac{1}{2}d_{ij}^2 \tag{9.3}$$

then, center this matrix as follows

$$B = HAH \quad \text{with } H = I - \frac{1}{L}\mathbf{1}\mathbf{1}^T \tag{9.4}$$

with $\mathbf{1} = [1\ 1\ 1\ \dots\ 1]^T$ a row of L ones, I, the identity matrix of dimension L. B can also be written as:

$$B = (HX)(HX)^T \quad \text{of size L} \times \text{L} \tag{9.5}$$

Consider now the eigenvalue decomposition of B as:

$$B = V_B \Lambda_B V_B^T \tag{9.6}$$

In our case, $L \ll N$ and the distance is Euclidean, hence all eigenvalues are positive. We can now reconstruct (map onto a location in Cartesian space) any $\mathbf{x}$ in X in any dimension from a minimum of one dimension up to a maximum of L dimensions, by considering that

$$B = (HX)(HX)^T = V_B \Lambda_B V_B^T \Rightarrow X = V_B \Lambda_B^{1/2} \Rightarrow X_d = V_{B,d} \Lambda_{B,d}^{1/2} : X \overset{\text{MDS}}{\mapsto} X_d \tag{9.7}$$

if we take the d largest eigenvalues. $V_{B,d}$ contains the eigenvectors belonging to the d

largest eigenvalues contained in the diagonal matrix $\Lambda_{B,d}$. The solution X_d retained by MDS is such that the mapped locations contained in the matrix

$$X_d = \begin{bmatrix} \mathbf{x}_{1,d}\, \mathbf{x}_{2,d} \ldots \mathbf{x}_{L,d} \end{bmatrix}^T$$

have their centroid as origin and the axes are chosen as the principal axes of X (proof not given here). The length of each vector $\mathbf{x}_{i,d}$ is d, the chosen mapping dimension. Traditional MDS was developed for Euclidean distances, but as an extension the same operations can be done on any distance matrix.

Consider the example illustrated in Figure 9.3. 1000 Earth models $\mathbf{x}_i$, $i = 1, \ldots, 1000$ are generated from a sequential Gaussian simulation (size $N = 10\,000 = 100 \times 100$; see Chapter 7) using a spherical anisotropic variogram model and standard Gaussian histogram. The Euclidean distance is calculated between any two models resulting in a 1000×1000 distance matrix. A 2D mapping ($d = 2$) is retained of the Earth models in Figure 9.3. What is important in this plot is that the (2D) Euclidean distance in Figure 9.2 is a good approximation of the (ND) Euclidean distance between the models. Each point i in this plot has two coordinates which are equal to:

$$\mathbf{x}_{i,d=2} = (v_{1,i}\sqrt{\lambda_1},\, v_{2,i}\sqrt{\lambda_2})$$

with $v_{1,i}$ the ith entry of the first eigenvector. However, the actual axis values are not of any relevance; it is the relative position of locations that matters because this reflects the difference between the Earth models. So, in all of what follows in this book, any axis

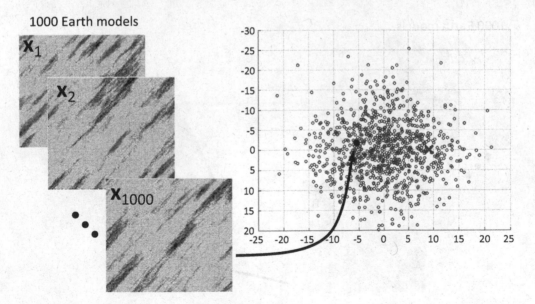

Figure 9.3 1000 variogram-based Earth models and their locations after projection with MDS: Euclidean distance case.

values are not shown in order to emphasize that only the relative location of points is what matters. Notice how the cloud is centered around $\mathbf{0} = (0,0)$ as was achieved through the centering operation above. Projecting models with MDS rarely requires dimensions of five or higher such that the Euclidean distance between locations in the map created with MDS correlates well with the actual distance specified.

Consider now an alternative distance definition between any two models. Using the same models as previously, two points are located (A and B; Figure 9.4). A measure of connectivity (the details of how this measure is exactly calculated are not given here) is calculated for each Earth model. Such measure simply states how well the high values form a connected path between those two locations. The distance is then simply the difference in connectivity between two models. Using this distance, we produce an equivalent 2D map of the same models (Figure 9.4). Note the difference between Figure 9.3 and Figure 9.4, although both plot locations of the same set of models. If the connectivity-based projections (Figure 9.5) are investigated, it is noted how the Earth models of the left-most group are disconnected (a lack of red color between location A and B), while models on the right reflect connected ones. However, any two models that map closely may look, at least by visual inspection, quite different.

Consider now the case where these models are used to assess uncertainty of a contaminant traveling from the source A to a well at location B. Suppose that we are interested in the arrival time of such contaminant, then Figure 9.6 demonstrates clearly that a connectivity distance nicely sorts models in a low dimensional space, while models projected based on the Euclidean distance do not sort well at all. This will be important later, when we attempt select Earth models based on travel times (or any other nonlinear response).

1000 Earth models

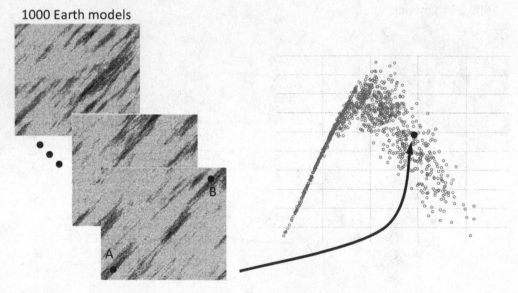

Figure 9.4   1000 Gaussian Earth models and their locations after projection with MDS: connectivity distance case.

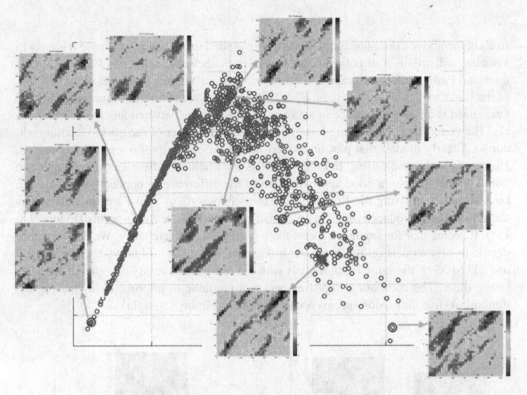

Figure 9.5 Location of a few selected Earth models.

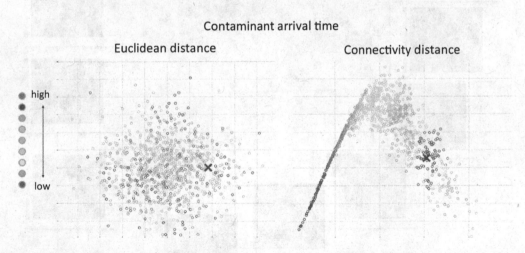

Figure 9.6 Plotting a response function (arrival time) at locations of models projected with different distances.

9.3.2 Determining the Dimension of Projection

In all of the above examples, Earth models were plotted in 2D, simply because it is easy to visualize a distribution of points in 2D. Consider another case, shown in Figure 9.7. One thousand Earth models are created, each showing a distribution of channel ribbons. Some of the Earth models as well as their 2D MDS projection are shown in Figure 9.6. The distance used is again a difference in measure of connectivity between any two Earth models. The connectivity is measured from the bottom-left corner of the grid to the top-right corner. Clearly models that plot on the left-hand side of Figure 9.6 are well connected. How good is this projection? In other words, is the Euclidean distance between any two points in this 2D plot a good approximation of the difference in connectivity? This can be assessed by plotting the Euclidean distances between any two Earth models versus the difference in connectivity. This plot will contain 1000×1000 points (Figure 9.8, left). It seems that for small distances, there is still some discrepancy. We can therefore decide to plot models in 3D, as shown in Figure 9.8 (middle). Figure 9.9 compares 2D and 3D MDS of the same Earth models with the same connectivity distance defined between them. The correlation improves, and will continue to improve by increasing the dimension (five dimensions seems accurate enough; Figure 9.8, right).

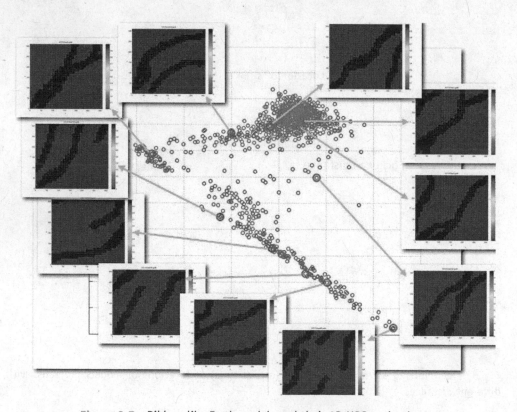

Figure 9.7   Ribbon-like Earth models and their 2D MDS projection.

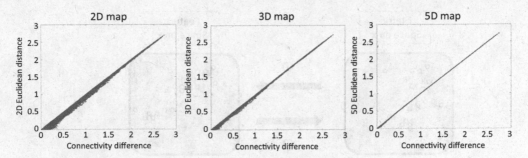

Figure 9.8 Plotting the Euclidean distance of 2D, 3D and 5D projection versus the connectivity difference.

9.3.3 Kernels and Feature Space

Defining a distance on Earth models and projecting them in a low dimensional (2D or 3D) Cartesian space presents a simple but powerful diagnostic on model variability in terms of the application at hand, at least if the chosen distance is reflective of the difference in studied response. Clearly, how model uncertainty is looked at is application dependent (see the difference between Figures 9.2 and 9.3). In Chapter 10, these plots will be used to select a few Earth models that are representative for the entire set (for target response uncertainty evaluation) and to assess what parameters are most influencing the response (sensitivity or effect analysis). However, as shown in Figure 9.9, the cloud of models in a 2D or 3D projection Cartesian space may have a complex shape, making the selection of representative models (as done Chapter 10) difficult. In computer science kernel techniques are used to transform from one metric space into a new metric space such that, after projecting in 2D, 3D and so on, the cloud of models displays a simpler arrangement.

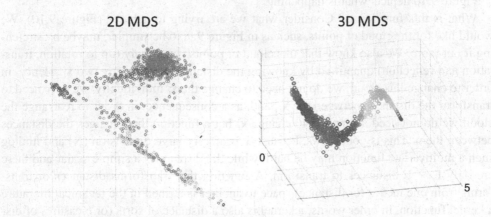

Figure 9.9 A 2D and 3D MDS map of the Earth models shown in Figure 9.6. The color indicates the connectivity of each model (blue = low, red = high).

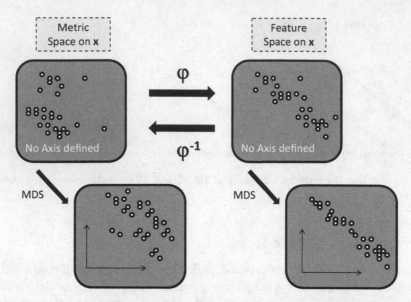

Figure 9.10   Concept of metric and feature space and their projection with MDS.

The goal, therefore, is to transform (change) the Earth models such that they arrange in simpler fashion in the MDS plot. To achieve this, consider an Earth model $\mathbf{x}_i$ and its transformation using some multivariate function φ:

$$\mathbf{x}_i \mapsto \varphi(\mathbf{x}_i) \quad \text{or} \quad \begin{pmatrix} \mathbf{x}_1 \\ \vdots \\ \mathbf{x}_L \end{pmatrix} \mapsto \begin{pmatrix} \varphi(\mathbf{x}_1) \\ \vdots \\ \varphi(\mathbf{x}_L) \end{pmatrix}$$

Figure 9.10 depicts what is happening.

What is this function φ? Consider what we are trying to achieve (Figure 9.10). We would like for the cloud of points, such as in Figure 9.9, to be simpler, maybe by stretching it out more. We also know that this cloud of points is uniquely (up to rotation, translation and reflection) quantified by knowing the distance between any two $\mathbf{x}_i$. Hence, in order to change this cloud, we do not need to change the $\mathbf{x}_i$ individually, we only need to transform the distances between the $\mathbf{x}_i$, and, as a consequence, in order to rearrange the cloud we do not need a function to change $\mathbf{x}_i$ but a function that changes the distances between them. This is good news, because $\mathbf{x}$ is of very large dimension (N) and finding such a multivariate function may be hard, while the distance is a simple scalar and there are only L × L distances to transform. A function that transforms distances or transforms from one metric (or distance) space to another is termed in the technical literature a kernel function. In other words, a kernel is also a distance of sorts (or measures of dissimilarity, but technically mathematicians call it a dot-product; the values of matrix B in Equation 9.5 are also dot-products). There are many kernel functions (as there are many distances and dot-products); we will just discuss one that is convenient here, namely the radial basis kernel (RBF), which in all generality is given by:

$$K_{ij} = k(\mathbf{x}_i, \mathbf{x}_j) = \exp\left(-\frac{(\mathbf{x}_i - \mathbf{x}_j)^T(\mathbf{x}_i - \mathbf{x}_j)}{2\sigma^2}\right)$$

Note that this RBF is function (a single scalar) of two Earth models $\mathbf{x}_i$ and $\mathbf{x}_j$ and of parameter σ, called the bandwidth, which needs to be chosen by the modeler. The bandwidth acts like a scalar of length. If σ is very large, then K_{ij} will be zero except when $i = j$, meaning that all $\mathbf{x}_i$ tend to be very dissimilar; while if σ is close to zero, then all $\mathbf{x}_i$ are deemed very similar. So it is necessary to choose a σ that is representative of the difference between the various $\mathbf{x}_i$. Practitioners have found that choosing the bandwidth equal to the standard deviation of values in the $L \times L$ distance matrix is a reasonable choice.

Note that the RBF is function of the Euclidean distance $(\mathbf{x}_i - \mathbf{x}_j)^T(\mathbf{x}_i - \mathbf{x}_j)$. We want to make this RBF kernel a bit more general, that is, by making it function of any distance, not just the Euclidean distance. This can be done simply by means of MDS as follows:

1 Specify any distance $d(\mathbf{x}_i, \mathbf{x}_j)$.

2 Use MDS to plot the locations in a low dimension (e.g., 2D or 3D), call these locations $\mathbf{x}_{d,i}$ and $\mathbf{x}_{d,j}$ with d the dimension of the MDS plot.

3 Calculate the Euclidean distance between $\mathbf{x}_{d,i}$ and $\mathbf{x}_{d,j}$.

4 Calculate the kernel function with given σ.

$$K_{ij} = k(\mathbf{x}_i, \mathbf{x}_j) = \exp\left(-\frac{(\mathbf{x}_{d,i} - \mathbf{x}_{d,j})^T(\mathbf{x}_{d,i} - \mathbf{x}_{d,j})}{2\sigma^2}\right)$$

Since we have a new indicator of "distance," namely K_{ij}, we also have a new "metric space', which is traditionally called the "feature space" (Figure 9.6). Now, the same MDS operation can be applied to matrix K. The eigenvalue decomposition of K is calculated and projections can be mapped in any dimension (Figure 9.9), for example in 2D using

$$\Phi_{f=2} = V_{K, f=2}\Lambda_{K, f=2}^{1/2}$$

where $V_{K, f=2}$ contains the eigenvectors of K belonging to the two largest eigenvalues of K contained in the diagonal matrix $\Lambda_{K, f=2}$. An illustrative example is provided in Figure 9.11. One thousand Earth models were mapped into 2D Cartesian space (the same as Figure 9.6). Shown on the right-hand side of Figure 9.11 are the 2D projections of models in feature space. Note how the complex cloud of Earth model locations has become more "stretched out" (left-hand side). In the next chapter we will see that this gives way to better model selection and quantification of response uncertainty.

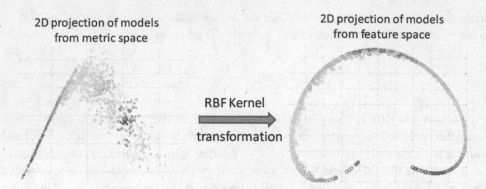

Figure 9.11 Comparison of locations of models and response function evaluation after projection with MDS from metric space and feature space.

9.3.4 Visualizing the Data–Model Relationship

The above application of MDS demonstrates how the possibly complex uncertainty represented by many alternative Earth models can be visualized through simple 2D or 3D scatter plots. In many applications, data are used to constrain models of uncertainty, as was discussed in Chapters 7 and 8. In other applications, new data become available over time and the current model of uncertainty needs to be updated. In a Bayesian context this means that the current "posterior," that is, the current set of models matching the data as well as reflecting the current prior information available, now become the "new prior."

It is, therefore, of considerable interest to study the relationship between any prior model of uncertainty and the data available. In Chapter 7 it was discussed that there may be a conflict between the prior and the data–model relationship (the likelihood), since they are often specified by different modelers. Can a simple plot be created, therefore, comparing the prior model uncertainty with the data?

To illustrate that this is indeed possible, consider a realistic case example. A synthetic but realistic reservoir case study, known as the "Brugge data set," named after the Flemish town where the conference involving this data set was held, is used as demonstration. As prior uncertainty, a number of input modeling parameters are unknown:

- The models are either binary rock types (sand vs background) each with different permeability and porosity characteristics or the system is directly modeled using the continuous permeability and porosity variables. In either case, spatial uncertainty is present.

- The geological scenario is uncertain: either the system contains sand fluvial channel objects within a background (binary) or the (binary) system was modeled using variogram-based techniques. However, the Boolean model was taken as deterministic, so was the variogram.

- The proportion of the sand rock type in the reservoir is uncertain with given probability distribution.

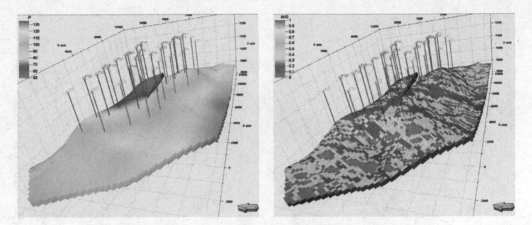

Figure 9.12 Pressure obtained by forward simulation (left), An example Earth model, the red color indicates channel sands, the purple indicates more shaly rocks (right).

An example Earth models is shown in Figure 9.12. All models are constrained to some "hard data," which are available from the wells drilled in the reservoir. These models represent the prior uncertainty.

The field data concern the historical monthly values of water rate, oil rate and pressure measurements in 20 producing wells over a 10 year period. The forward model is a (finite difference) simulation model of flow in the subsurface. The initial and boundary conditions are assumed known, as are many of the fluid properties. Figure 9.12 shows the pressure variation in this reservoir obtained by forward simulation on the model on the left.

To visualize prior uncertainty as well as the historical production data we proceed as follows. Firstly, apply the forward model g to each of the L Earth models $\mathbf{m}_i$, resulting in a forward model response $\mathbf{g}_i = g(\mathbf{m}_i)$, which is a vector containing the time series production responses of oil/water rate and pressure. Figure 9.13 shows these responses for one of the 20 producing wells. As distance simply define some distance between the forward model responses for each of the models:

$$d(\mathbf{m}_i, \mathbf{m}_j) = d_g(g(\mathbf{m}_i), g(\mathbf{m}_j)) \text{ for example } d_g(g(\mathbf{m}_i), g(\mathbf{m}_j)) = \sqrt{(\mathbf{g}_i - \mathbf{g}_j)^T(\mathbf{g}_i - \mathbf{g}_j)}$$

In addition, we have the real data from the actual field, which we term $\mathbf{d}$. In fact, these data can be seen as the response from the "true" Earth $\mathbf{m}_{\text{true}}$

$$\mathbf{d} = g(\mathbf{m}_{\text{true}})$$

if we assume that the forward model reflects correctly the Earth's response and an Earth model $\mathbf{m}$ captures the true Earth realistically. This means that we can also calculate the

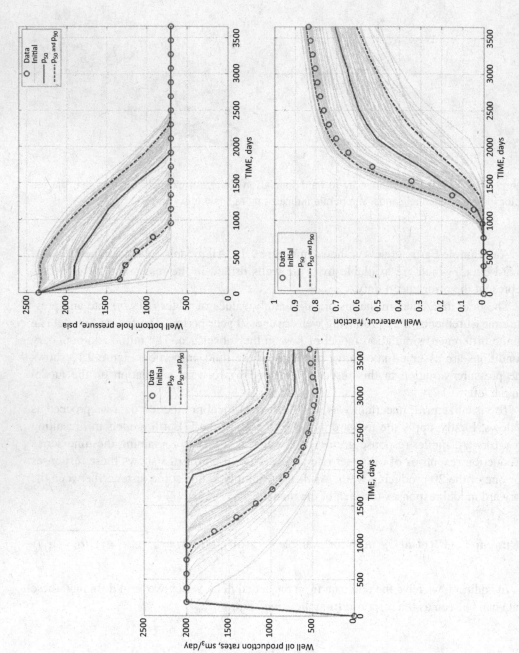

Figure 9.13   Oil production rate over time (left), well-bore pressure (top right), fraction of production that is water (bottom right).

distance between the responses of the models $\mathbf{m}_i$ and the true Earth as represented by the data, namely:

$$d_g(g(\mathbf{m}_i), \mathbf{d}) = \sqrt{(\mathbf{g}_i - \mathbf{d})^T (\mathbf{g}_i - \mathbf{d})}$$

This means that we have an L + 1 (L Earth models and 1 true Earth) distance matrix that allows plotting, using MDS, the L models as well as the true Earth. Figure 9.14 shows the result for the Brugge data set. One observes that in this case, the "true Earth" (or better its response) plots with the various alternative prior Earth models. This is good news: it means that the modeling effort at least captures what the true unknown is, in terms of this specific response. Should the true Earth plot be outside of this scatter of points, as is hypothetically shown in Figure 9.14, then either the prior uncertainty is incorrect, the data are incorrect or the data–model relationship is incorrect. Such evaluation is important

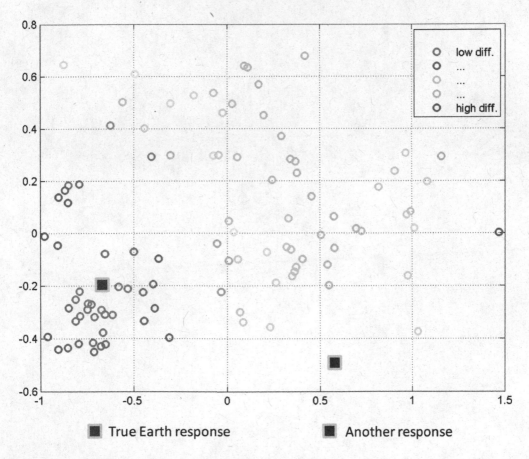

Figure 9.14 MDS plot of the response of 65 Earth models as well the response from the true Earth (field data). The color indicates the difference (diff.) in response with the field data.

before addressing the issue of integrating these data into the Earth models, for example using inverse modeling techniques as discussed in Chapter 7.

Further Reading

Borg, I. and Groenen, P. (1997) *Modern Multidimensional Scaling: Theory and Applications*, Springer, New York.

Peters, L., Arts, R.J., Brouwer, G.K. *et al.* (2010) Results of the Brugge benchmark study for flooding optimization and history matching. *SPE Reservoir Evaluation & Engineering*, **13**(3), 391–405. SPE-119094-PA. doi: 10.2118/119094-PA.

Schöelkopf, B. and Smola, A. (2002) *Learning with Kernels*, MIT Press, Cambridge, MA.

10
Modeling Response Uncertainty

Modeling response uncertainty should not be considered as an afterthought, for example as a simple sensitivity analysis performed on the parameters of a single (deterministic) Earth model from which a single prediction was made. Instead, a full exploration of the range of uncertainty of responses based on multiple alternative Earth models, keeping the intended application always in perspective, is desired.

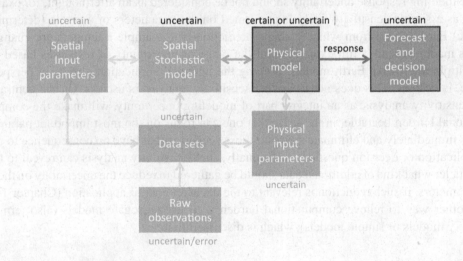

10.1 Introduction

In the previous chapter the various CPU challenges with modeling uncertainty in the Earth Sciences were discussed. It was discovered how evaluating a transfer function on each Earth model to assess some response uncertainty is impossible in most practical cases. Instead, a few representative models need to be selected. Often, one would be interested in only three models: one model with a typically "low" response (however low is defined, for example, low predicted contamination), a typical or median response, and

Modeling Uncertainty in the Earth Sciences, First Edition. Jef Caers.
© 2011 John Wiley & Sons, Ltd. Published 2011 by John Wiley & Sons, Ltd.

a high response. We can make such a selection more statistically motivated by picking the Earth model whose response is such that 10% of the model responses are lower than the model response of this so-called P10 model (lowest decile). In similar ways we can then define the median model as the Earth model such that 50% of models have smaller response and 50% have higher response, namely the P50 model. The P90 model is then the antipode of the P10 model. For example, if the goal were to calculate the P10, P50 and P90 of the cumulative oil production of an oil field, then one would ideally only need to pick the correct three reservoir models and perform flow simulation on them.

Picking the "correct" Earth models would be easy if the relationship between Earth model and response function output was linear. Since in reality this relationship is highly nonlinear, selecting which model to evaluate the response function is not a trivial task. Some techniques for selecting representative models are discussed in this chapter. In line with the philosophy of this book, the selection of models should be decision or application driven. Indeed, one would not select the same three climate models to explore different hypotheses about the detailed, regional nature of the global climate dynamics, the causes of observed patterns of climate change, the generation of regional forecasts, or just to predict the global mean temperature change.

Modeling response uncertainty should not be considered as an afterthought, for example as a simple sensitivity analysis performed on the parameters of a single (deterministic) Earth model from which a single prediction (for example a temperature change) was made. Instead, a full exploration of the range of uncertainty of responses based on multiple alternative Earth models, keeping the intended application always in perspective, is desired. This does not mean that sensitivity analysis is useless. On the contrary, a sensitivity analysis, as an integral part of modeling uncertainty, will make the computational burden bearable, in the sense that one can focus on the most important parameters immediately and eliminate studying those scenarios that have no consequence to the application or decision question. Additionally, such sensitivity analysis can reveal to the modeler what kind of additional data should be gathered to reduce the uncertainty on these parameters, if such reduction is relevant to the decision goal or application (Chapter 11). Another way to relieve computational burden is to use surrogate models (also termed proxy models or simple models), which is discussed first.

10.2 Surrogate Models and Ranking

If evaluating the response function or simulation model on an Earth model is too expensive, then one option is simply to use a less expensive (less CPU demanding) function that approximates the original or "full" response function. Such an approximating function is called a "proxy" function, "surrogate model" or "simple model". The surrogate or proxy model mimics the behavior of the full simulation model. A simply analogy would be to approximate $\sin(x)$ by x which is easy to evaluate but is only a valid approximation when x is small. In similar ways, surrogate models will approximate the full model well for certain types of conditions or ranges of physical and spatial parameters or even for certain types of applications. The surrogate model can be a simpler physical model, an analytical approximation of a more complex partial differential equation, but can also be an interpolation model that interpolates between points evaluated on the full simulation model. An example of the latter approach is response surface analysis, which is discussed next.

One common usage of proxy functions is to use ranking to guide the selection of Earth models. The central principle behind ranking is to use some simple measure to rank Earth models from lowest to highest as defined by the surrogate model and then run a full transfer function on selected Earth models to assess, for example, a P10, P50 and P90 response or whatever else is desired. If 100 Earth models are created then one would only need to apply the full simulation on the 10th, 50th and 90th Earth model as ranked by the proxy function. This would define a model of uncertainty without performing a large number of response evaluations. This works well as long as a good ranking measure (or surrogate model) can be found, since the method is sensitive to the degree of association between the surrogate model and the actual response function. This technique often underperforms with regard to other techniques. Nevertheless, it is a simple and, in many cases, useful technique.

10.3 Experimental Design and Response Surface Analysis

10.3.1 Introduction

A widely used approach in statistics for evaluating response variation is the experimental design (ED) technique in combination with response surface analysis. These are general statistical techniques, which are applied in many areas of sciences, not just Earth Sciences, such as opinion polls, statistical surveys, but often refer to controlled experiments. In general, in ED, the "experimenter" (in our case the modeler) is interested in the effect of some process (the "treatment") on some objects (the "experimental units"), which may be people, plants, animals, and so on. In our case the modeler is interested in the effect of some parameters on some response. To evaluate such impact, one could generate many sets of parameters and evaluate their effect on the response. If such parameter sets are chosen randomly, then we have Monte Carlo simulation, which is too CPU demanding. However, it is possible to choose the parameters more carefully, and "design" the experiment better (than random), then evaluate the responses for this designed experiment (parameter choices). Finally, a mathematical function, termed a "response surface," can be fitted through these response values. This function can then be used as a fast approximation (a surrogate model) to the true response for any set of parameter values. In doing so, it is necessary to define a relationship between the response of interest and a set of uncertain parameters. This relationship is often a simple linear or second order polynomial function of the uncertain parameters with the response. The aim of an experimental design is to define the minimum number of parameter sets in order to obtain the best fit to the response surface.

10.3.2 The Design of Experiments

The first step in response surface analysis is to choose the design of the experiment, that is, what combination of parameters is selected to evaluate the response function on. Ideally, if we could choose that set of parameter combinations we know most influences

factor		levels indicator		Interaction	response
X (temp)	Y (ratio)	A	B	AB	Z (strength, MPa)
300	1	-1	-1	1	9
500	1	1	-1	-1	7
300	9	-1	1	-1	5
500	9	1	1	1	9.5

Figure 10.1 Example case for a 2^2 full factorial design.

the response, then we would design our experiment as such. The problem is that we often don't know this a priori (ahead of time). In fact, finding the combination of parameters that has most influence is often a goal on its own. The experimental design literature offers various designs, which have been tested for many applications and for which some knowledge has been gathered, either theoretically or experimentally on how well they work. This is a field on its own and only a few designs that are relevant are discussed here to explain what is being done.

In ED, a parameter would be called a "factor." Typically, these factors are then discretized into s levels, for example in a two-level design one determines for each factor what is a high value and low value; this is a decision made by the modeler. In a so-called full factorial design with k factors and s levels, one chooses s^k experiments, as for example illustrated in Figure 10.1. In the case of a 2^2 factorial design, there are four parameter combinations for which the response function is evaluated. Consider a practical example of testing the rock strength (Z, e.g., strength in simple tension) as function of temperature (X, in F) and sand/shale ratio (Y). Firstly, two levels for each factor are chosen; for temperature: 300 (low) and 500 (high), for the shale/sand ratio: 1 (low = more shale) and 9 (high = more sand). Each level is first coded into a -1 (low) and a 1 (high) (Figure 10.1); the so-called interaction term XY (how temperature and shale/sand ratio jointly influence the strength) is simply the product of the two. If we are interested in how much the "effect" is of X on the tensile strength then we can simply take the average response value at the higher of the two temperatures (associated with 1) and subtract the average at the lower two temperatures, that is:

$$\text{Estimate of effect } X = \frac{7 + 9.5}{2} - \frac{9 + 5}{2} = 1.25$$

Similarly:

$$\text{Estimate of effect } Y = \frac{5 + 9.5}{2} - \frac{9 + 7}{2} = -0.75$$

In similar vein, we can do this for the interaction term and take the response values associated with 1 and -1 and make a similar calculation:

$$\text{Estimate of effect } XY = \frac{9 + 9.5}{2} - \frac{7 + 5}{2} = 3.25$$

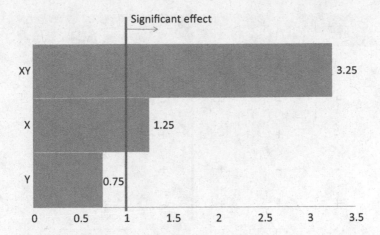

Figure 10.2 Pareto plot listing the effects from large to small. The red line indicates the level above which effects are considered significant.

In this simple example, there are basically four "treatment combinations" (combinations of plus ones and minus ones) and one measurement for each treatment; hence we have a perfectly designed experiment, assuming there is no measurement error. In the case of measurement error, it would be necessary to replicate the same treatment combination to estimate the error, and the estimates of the effect will then be prone to error.

Often the effects are listed in what is termed a Pareto chart (Figure 10.2), which list the magnitudes of each effect from large to small. This allows the modeler to focus immediately on the most important effects to include in further analysis (in this case, for example, the X and XY effects). In this Pareto chart a significance level is also displayed, that is, the level above which effects are considered significant. There exist statistical techniques for determining this level but this is outside the scope of this book. In most practical cases, one simply retains those effects that are clearly larger than the others.

Another way to get the effects is to basically fit a surface to the data:

$$Z = b_0 + b_1 X + b_2 Y + b_3 XY$$

Notice again that we need to determine four coefficients from four data points in a 2^2 design; hence a perfect fit is obtained (four equations with four unknowns). A simple relationship exists between the coefficients and the estimate of the effects, namely:

Regression coefficients b_1 = half the effect estimate of X ,

This is similar for other coefficients.

The term 2^k increases rapidly when the number of factors k increases. For example, when studying 10 parameters the number of treatment combinations is 1024, and this will increase even more when the number of levels s increases; hence only a fraction of treatment combinations can be studied. This is termed a fractional factorial design. In a general form a two-level fractional factorial design is written as 2^{k-p}; an increasing value

First Fraction			Second Fraction		
A	B	C	A	B	C
+	−	−	−	−	−
−	+	−	+	+	−
−	−	+	+	−	+
+	+	+	−	+	+

| A, B, C, ABC | 1, AB, AC, BC |

Figure 10.3 Example of a fractional design for 2^{3-1}. A, B and C are the three factors; two levels (− and +).

of p means that lesser treatment combinations are studied. The fraction of the total possible treatments is thus equal to $(^1\!/_2)^p$. The question then is: what fraction is retained, that is, what effects do we focus on and group into a single fractional design?" One may want to focus on main effects or one may want to focus on interaction effects. For example in 2^{3-1} design two designs can be created where $^1\!/_2$ the fraction is used for each design. Figure 10.3 provides a logical choice where one fraction focuses on the main effects and the other on the interactions. Figure 10.3 can be logically understood as follows: the first fraction studies the effects whereby AB = C (for example $(+)(−) = (−)$), or ABC = +1, and the second fraction studies the effects whereby AB = −C (for example $(−)(−) = −(−)$) or ABC = −1. It can be stated that the first fraction studies only a positive effect of ABC while the second studies a negative effect. A similar split can be made when k increases.

10.3.3 Response Surface Designs

The response surface is, as the name suggests, a tool for fitting a surface to a set of data. A true surface exists, namely, the true relationship between the parameters/factors and the response. The estimation or fitting procedure will yield a surface that is different from the actual surface. An important element of this methodology is the response surface design. A design that is often used and a natural extension of previously discussed designs are the central composite designs or CCDs. The basic CCD (as many exist) consists of three parts:

- A fractional or full factorial design.

- Axial points.

- Central points.

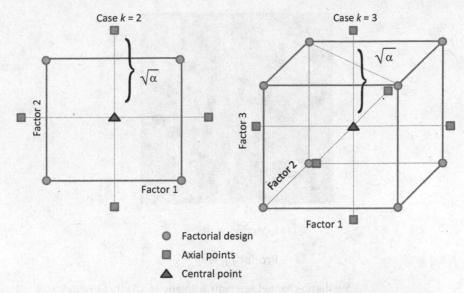

Figure 10.4   Example of CCD for the cases $k = 2$ and $k = 3$.

Figure 10.4 shows a basic CCD with $k = 2$ and $k = 3$. The term $\sqrt{\alpha}$ determines how far the axial points lie from the other design points, typical values used are $\alpha = 2$ or $\alpha = 1$. In terms of number of design points therefore:

$$k = 2 : 4 + 4 + 1 = 9 \text{ design points}$$
$$k = 3 : 8 + 6 + 1 = 15 \text{ design points}$$

10.3.4 Simple Illustrative Example

The response surface methodology can be illustrated using a simple example. A channel Earth model with known channel locations (Figure 10.5) has two uncertain input parameters, the (constant) permeability (how well the channel conducts flow) and porosity (how much volume there is to flow through) of the channel. The location of wells is shown in Figure 10.5. Some wells inject water into the system to drive out the oil towards the producing wells. The goal is to model the uncertainty of the oil production rate over time for the entire area, that is, from the three producing wells.

A central composite design with $\alpha = 2$ is used (Figure 10.6). Since there are two factors only, we have a design with nine points. Response evaluations for all nine models are performed (Figure 10.6). From those response evaluations, a relationship is built using a second order polynomial to describe the oil production rate (at a given time) versus these two parameters. The fitted response surface is also compared with an exhaustively evaluated response surface by doing $25 \times 25 = 500$ response evaluations. The fitted response surface compares favorably with the true response; note that both are smooth surfaces.

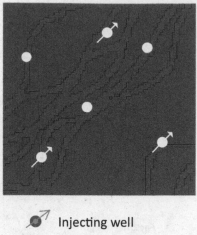

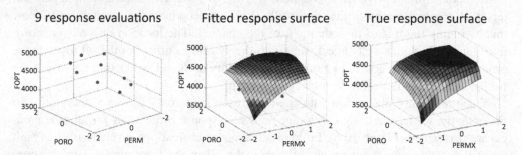

Figure 10.5   Synthetic channel reservoir with known channel geometry.

factor 1 (poro)	factor 2 (perm)	actual poro	actual perm	response
-1	-1	0.15	250	4269.135
-1	1	0.15	750	4607.682
1	-1	0.3	250	4772.588
1	1	0.3	750	4800
-1.41421	0	0.1189	500	4260.636
1.41421	0	0.3311	500	4800
0	-1.41421	0.225	146.4466	4205.167
0	1.41421	0.225	853.5534	4771.403
0	0	0.225	500	4771.403

Figure 10.6   Experimental design and response surface for the channel case.

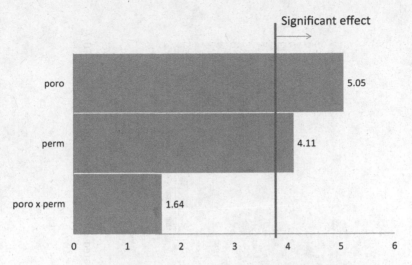

Figure 10.7   Pareto plot for the channel case.

As outlined above, the methodology allows quantifying which factors have the most effect, that is, highest sensitivity with regard to the response. The Pareto plot of the two factors as well as their interaction effect is plotted in Figure 10.7. Clearly, the interaction between porosity and permeability in estimating oil production is negligible compared to the main effects.

To assess the uncertainty of the oil production rate it is now assumed that the response surface is a good approximation for the actual flow simulation. Various porosity and permeability values are randomly sampled, each according to a uniform distribution (any other distribution model could have been used) (Figure 10.8). Simply by evaluating the response surface for each pair of the sampled porosity and permeability values, the histogram of oil production rate in Figure 10.8 is obtained as a representation of oil production rate. This histogram is a representation or model of uncertainty for the oil production rate.

10.3.5 Limitations

As mentioned, the response surface is not equal to the true relation between the response and the parameters because the response cannot be exhaustively evaluated, a fractional design of some sort needs to be chosen by the modeler. That inequality manifests itself on two levels. Firstly, the accuracy of the response surface depends very much on the chosen fractional design, whether it is a CCD or any other available in the literature. If the design is such that the main fluctuations in the response are missed than large errors may occur. Secondly, the response surface is fitted by minimizing the variation of the error between the surface and the response values (in statistics called an error variance). This necessarily leads to a smooth surface (RSM is very much like Kriging (Chapter 6), it generates a smooth representation of a more variable truth). The latter observation has several consequences.

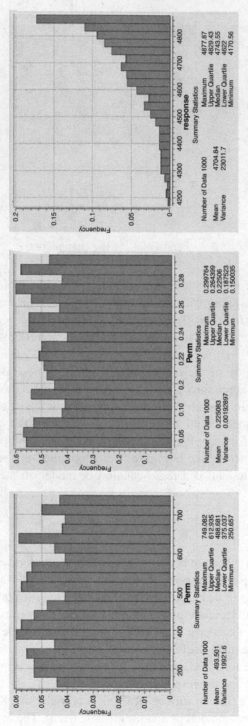

Figure 10.8 Histogram of sample porosity and permeability. Histogram of the response evaluated with the response surface in Figure 10.7.

Firstly, the response surface methodology performs well when the response varies smoothly with regard to the input parameters and when the variability of the input parameters can be represented by choosing a few levels (e.g., three levels). As a consequence, when the true response is highly uncertain, that is, varies a lot with regard the input parameters, the methodology may understate the response uncertainty. Choosing a few limited levels on the parameters may also lead to a false sense of security because, in actuality, the response may vary due to a small change in parameter values that cannot be captured by a few coarsely chosen levels of experimental design. As a result, to be truly safe in not missing important response changes, a large number of response evaluations may be needed when performing experimental design. To know how many response evaluations are needed ahead of time (a priori) is difficult to determine.

Secondly, the response surface method does not work for parameters that have no natural levels (e.g., can be classified into low, medium, high). This is particularly the case for parameters that represent a scenario: such as a parameter that represents the choice of the training image, parameters that represent the choice of the physical model (or describes a part of it) or parameters that represent various interpretations such as for example of a structural model (Chapter 8). To be able to perform experimental design for any type of study and variable we will again rely on distances.

This technique is therefore a useful tool, for example using it as a proxy or surrogate model, but not it is not general enough to be a methodology on its own for modeling uncertainty in the Earth Sciences.

10.4 Distance Methods for Modeling Response Uncertainty

10.4.1 Introduction

In the previous chapter the notion of a distance (a single scalar value) was introduced to define uncertainty as a function of the practical decision questions or response evaluations. The purpose of this distance is to bring some structure (less randomness) into the large uncertainty represented by a set of alternative Earth models. This distance will now be used to select models, but in a way that is more effective than the ranking technique and more general than the experimental design technique.

Uncertainty as represented by a set of alternative Earth models is often by itself of little interest (Chapter 3): it is that part of uncertainty that impacts variability and uncertainty in the decision that matters, if the goal is to make decisions based on the models built. Discussed in Chapter 4 were various techniques for sensitivity analysis that allow the discovery of what mattered for the particular decision question at hand, as was done with the experimental design technique (for example using a Pareto plot as in Figure 10.2). In this chapter this question of sensitivity is returned to but now involving several Earth models as representations of uncertainty; this poses more challenges than the simple sensitivity problem (sensitivity to cost or prior probabilities for example) treated in Chapter 4. Hence, two challenges are being defined to be tackled by the distance approach: decision-driven selection of models and sensitivity of responses to input parameters (physical or spatial).

10.4.2 Earth Model Selection by Clustering

10.4.2.1 Introduction

Clustering techniques are well-known tools in computer science as well as in many other areas of science, including Earth Sciences. Two forms of clustering are known: supervised and unsupervised. We will primarily deal with the latter. In unsupervised clustering, the aim is to divide a set of "objects" into mutually exclusive classes based on information provided on that object. What is unknown are the number of classes and what features or attributes of the object should be used to make such division. For example, 100 bottles of wine (objects) are on the table, but the label for each bottle is hidden to the expert. A wine expert can group these bottles for example by grape variety or region of origin (attributes) simply by tasting the wine. The better the expert, the more refined the grouping will be and the more classes may exist. The decision on what attributes to use is therefore an important aspect of the clustering exercise and the topic of considerable research in computer science known as "pattern recognition." The combination of (grape variety, origin) is an example of a pattern. In our application, an Earth model is such an "object" and the aim is to group these Earth models into various classes with the idea that each class of models has a similar response, without the need for evaluating responses on each model. If this can be achieved successfully, then one single model of a cluster or group can be selected for response evaluation. Similarly, the wine expert can take a bottle out of each group to represent the variety of wines on the table without needing to select all 100 bottles of wine. The labels revealed for each bottle is the equivalent of our response function.

The number of classes, clusters or groups can either be decided on (a) how many response evaluations are affordable (a CPU issue) or (b) how many response evaluations are needed to obtain a realistic assessment of uncertainty on the response (an accuracy issue). This question is addressed later; addressed first is the question of how to cluster without evaluating responses on each Earth model, which would defeat the purpose of clustering itself.

If most clustering techniques in the computer science literature (e.g., k-means clustering, tree-methods, etc.) are considered then it is observed that the mathematics behind these methods calls for the definition of a distance. Indeed, a distance will define how similar each object is to any other object (recall our puzzle pieces analogy of similarity), allowing grouping of objects. However, the response function itself cannot be used to define such distance, a distance that is relatively easy and rapid to determine is needed. A wine expert could use the difference in wine color, color intensity, smell and difference in coating (or formation of "legs") on the glass as a way to distinguish wines without even tasting them (or worse, looking at the label). Similarly, for Earth models, the definition of a meaningful distance will make clustering effective and efficient. The elegance here lies in the fact one requires just a single distance definition, not requiring necessarily the specification of attributes or features to sort models, but one should be able to evaluate this distance rapidly. For this purpose, standard distances can be used, such as the Euclidean distance, Manhattan distance or Hausdorff distance, or surrogate/proxy models can be used. In using surrogate models, one would simply evaluate the response on each Earth model using the proxy function, then calculate some difference (e.g., a least square distance) between the surrogate model responses. The use of surrogate/proxy models will always be a bit field specific, namely if the purpose is flow simulation, then the field of reservoir engineering provides many fast approximate flow simulators (e.g., streamline simulation), including analytical solutions that can be used to calculate distances. In a more general application, one could create many fine scale models (possibly millions of cells), coarsen these models by some averaging procedure (upscaling), then use the responses of these coarse scale models to calculate distances. Or, one could use the response surface methodology as the definition of a surrogate. Note that the surrogate model is not used to approximate the response, but to approximate the difference (distance) in response, a subtle but important distinction.

10.4.2.2 k-Means Clustering

The computer science and statistical literature offer many clustering techniques. A simple technique is termed k-means clustering. The goal of the k-means clustering method is to cluster n objects into k classes. The value of k is specified by the modeler. In traditional k-means clustering, such objects are characterized by m attributes (e.g., fossils with given length and width). The objects are then plotted in m-D Cartesian space, as shown in Figure 10.9. The following algorithm summarizes the k-means algorithm, also summarized in Figure 10.9:

1 Initialize by choosing randomly m cluster centers.

2 Calculate the distance between each object and the cluster centers.

3 Assign objects to the center closest to it.

4 Calculate for each cluster a new mean (cluster center) based on the assigned objects.

5 Go to step 2 until no changes in the cluster means/centers are observed.

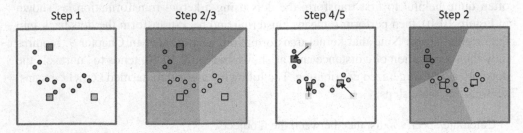

Figure 10.9 Steps in k-means clustering. In this simple case, convergence is after one iteration.

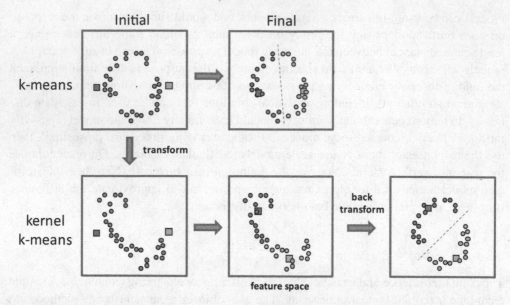

Figure 10.10 Comparison between *k*-means clustering and kernel *k*-means clustering.

The name of this clustering algorithm now becomes apparent. The *k* clusters are defined by cluster centers which are calculated as the mean of the objects attributes. Note that the number *k* needs to be specified and may not be easy to know a priori, particularly of the object is characterized by a large number of objects. The *k*-means clustering method does not necessarily require the specification of attributes; the above algorithm works as long as the distance between objects is known. A distance is a more general form of quantifying difference between objects than a difference in attributes of objects, which is actually a specific form of distance specification.

k-means works well for cases such as in Figure 10.9, but goes wrong in more difficult cases such as in Figure 10.10 where the variation of objects/points in the 2D plot is quite "nonlinear". Clearly two clusters exist in Figure 10.10. The *k*-means method is sensitive to the initialization of the algorithm in step 1. An unfortunate initialization leads to a clearly wrong grouping of objects. A solution that works well is to make the distribution of dots in this 2D plot more "linear" or at least more "aligned" in a specific direction. Chapter 9 presented the kernel transformation as a way to achieve this. Therefore, it is often quite helpful to first transform the dots using a kernel transformation, as shown in Figure 10.10, then perform *k*-means clustering and back-transform the dots back into the original space. Note that kernel transformation, as presented in Chapter 9, requires only the specification of a distance and, as shown in Figure 10.10, tends to "unravel" the dots into a more organized distribution. The following algorithm, termed kernel *k*-means, summarizes these steps:

1 Calculate/specify a distance between the n objects.

2 Transform the objects into a kernel/feature space.

3 Apply the *k*-means algorithm presented above.

4 Back-transform the results into the original Cartesian axis system.

10.4.2.3 Clustering of Earth Models for Response Uncertainty Evaluation

Figure 10.11 shows a summary of the clustering technique for model selection on a hypothetical example and is summarized as follows:

1 Create multiple alternative Earth models by varying several input parameters and generating several for each such fixed input (spatial uncertainty).

2 Select a distance relevant to the actual response difference.

3 Use this distance to create a projection of the Earth models using MDS.

4 Transform this set of points using the kernel transformations (Chapter 9).

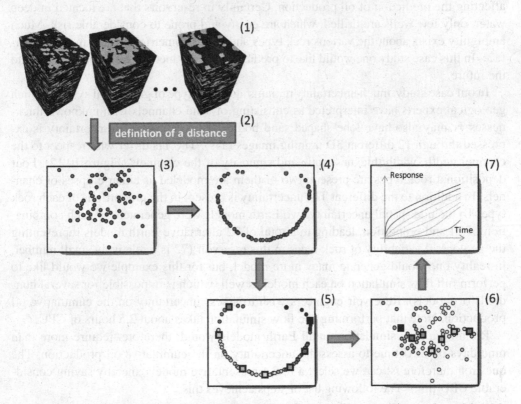

Figure 10.11   (1) Generation of Earth models, (2) defining a distance, (3) mapping with MDS, (4) transformation using kernels, (5) clustering using *k*-means and colored with the group to which they belong, (6) back to MDS mapping and identifying of Earth models for response evaluation, (7) responses.

5 Cluster this set of points into a group using clustering techniques such as k-means and find the cluster centers in the original MDS map.

6 Select the Earth models closest to the cluster centers.

7 Evaluate the response on the selected Earth models.

Note that this relies on the property of the original k-means (or kernel k-means) algorithm that only a distance is required to perform clustering. The clustering here is performed after projecting the high dimensional objects, namely the Earth models, into a low dimensional plot (for example a 2D plot such as in Figure 10.11). To make the practicality of this method apparent this technique is now applied to a computationally complex case study.

10.4.3 Oil Reservoir Case Study

Uncertainty about the geological system present in oil reservoirs is a common problem affecting the prediction of oil production. Certainly in reservoirs that are located in deep water, only few wells are drilled, which are costly and prone to considerable risk. Much ambiguity exists about the various rock types and their geometries present in the subsurface. In this case study one would like to predict the oil production over several years into the future.

In our case study, much uncertainty remains about the type of geological system, which geological experts have interpreted as consisting of sand channels with unknown thicknesses or may also have lobe-shaped sand bodies. This depositional uncertainty is expressed through 12 different 3D training images (TIs). The TIs differ with respect to the channel width, width/thickness ratio and sinuosity of the channels (Figure 10.12). Four depositional rock types are present; two of them are modeled as either ellipses or channels. In addition to the different TIs, uncertainty is present in the proportions of each rock type. To include spatial uncertainty, two Earth models were generated for each combination of TI and proportion, leading to a total of 72 alternative Earth models representing the geological variability of rock types in this reservoir (72 is a relatively small number, in reality one would generate many more model, but for this example we would like to perform full flow simulation on each model as well, which is impossible for several hundreds of models). Reservoir engineers wish to assess uncertainty on the cumulative oil production. Note that performing one flow simulation takes about 2.5 hours of CPU.

Running a flow simulation on all Earth models would, therefore, require more than nine days of CPU time to assess the uncertainty on the cumulative oil production. The question therefore is: can we select a few representative models, thereby saving considerable CPU time? The following list of steps achieves this:

1 Create 72 Earth models using the techniques covered in Chapter 6 (Figure 10.13).

2 Run a proxy (fast) flow simulator that approximates the physics of flow reasonably well (in our case a streamline simulator is used, which ignores the compressibility of fluids).

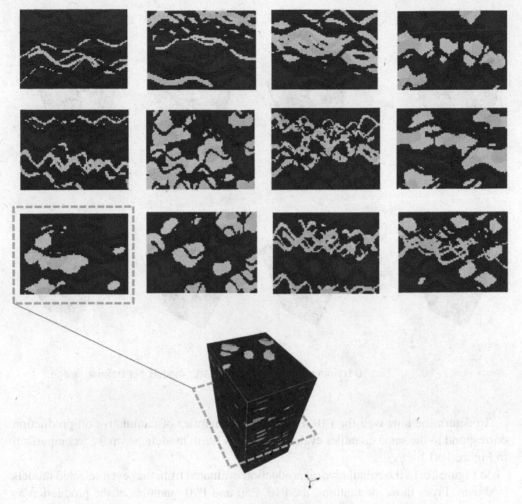

Figure 10.12   2D sections of a set of twelve alternative 3D training images.

3 Use the output response (cumulative oil) of this proxy simulator to calculate a distance between Earth models as the difference in the proxy simulators output response.

4 Map the Earth models in 2D and perform multidimensional scaling (Figure 10.14).

5 Cluster the Earth models into a limited number of groups, for example as many groups as we have time to run the full simulator (seven is chosen here).

6 Find the model closest to the cluster center (Figure 10.14).

7 Run the full simulation model on those models and calculate the lowest decile (P10), median (P50) and upper decile (P90) by interpolation (Chapter 2) from the seven evaluated responses (Figure 10.14).

Two example
training images   Three Earth Models

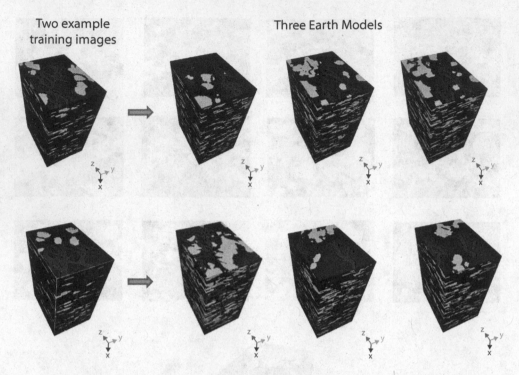

Figure 10.13   Two 3D training images and three Earth models per training image.

To determine how well the P10, P50 and P90 quantiles of cumulative oil production correspond to the same quantiles evaluated with all Earth models, we make a comparison in Figure 10.14.

In Figure 10.14 the cumulative oil production evaluated from the seven selected models is shown. From those simulations, the P10, P50 and P90 quantiles of the production as a function of the time can be computed. It is observed that the quantiles estimated from the seven models selected by clustering are very similar to the quantiles derived from the entire set of 72 models, which required ten times more computing time.

10.4.4 Sensitivity Analysis

Model selection using distance-based clustering can also provide insight into which geological or physical parameters are the most influential on the response. This information is often very helpful in determining what matters and what kind of data should be gathered to reduce uncertainty on the most important parameters. A very simple way of determining which parameter has most "effect" on the uncertainty of the response, is to do the same as in experimental design, except that now we do not use a standard design (which may work well on average for many applications) but use the models selected through clustering, which is more targeted towards the application at hand, that is, towards response uncertainty evaluation.

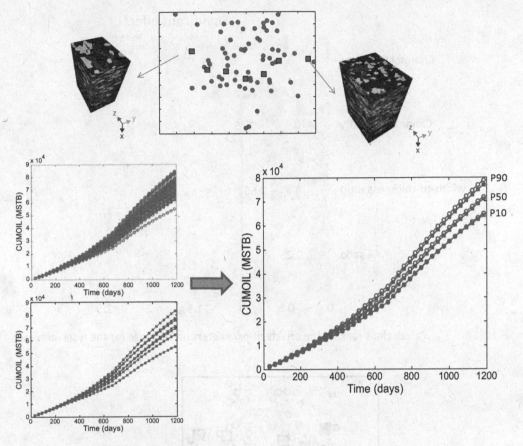

Figure 10.14 Top: MDS plot with the location of 72 Earth models (red circles) and the selected models closest to the cluster centers (blue squares). Bottom: response evaluation on the seven models (blue) and 72 exhaustive response evaluations with P10, P50 and P90 calculated for both sets.

Returning to the case study, then such a sensitivity study can be performed for the field cumulative oil production on four parameters: channel thickness, width/thickness ratio, channel sinuosity and percentage of sand (as defined by the three facies probability cubes). Figure 10.15 shows the sensitivity of each parameter on the cumulative oil production. The red line corresponds to a user chosen threshold, meaning that parameter values which cross the line are considered influential to the response. Clearly, the channel thickness is the most influential parameter for cumulative oil, followed by the channel sinuosity and the width/thickness ratio. The distance approach allows other ways of sensitivity analysis, not necessarily using response surface or experimental design techniques. The clustering of models provides a lot of information, since for each Earth model clustered in a group or cluster the generating parameters can be identified (Figure 10.16). Studying the joint variation of parameters within a cluster as well as the joint variation of parameters between clusters provides an insight into which combination of parameters

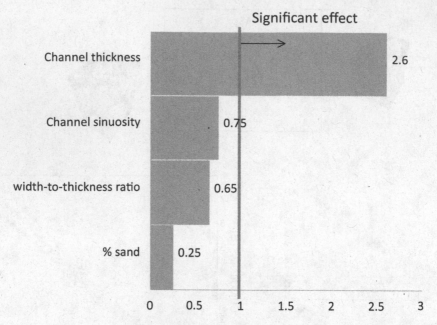

Figure 10.15   Pareto chart ranking the effects of parameters on response for the reservoir case.

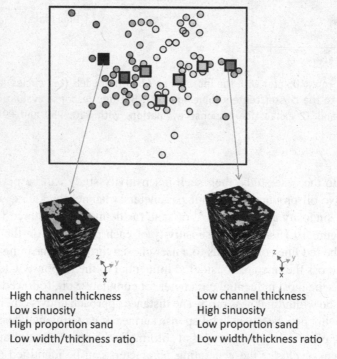

High channel thickness
Low sinuosity
High proportion sand
Low width/thickness ratio

Low channel thickness
High sinuosity
Low proportion sand
Low width/thickness ratio

Figure 10.16   Identifying generating parameters after clustering. Here the generating parameters of the cluster centers (the selected models) are shown.

is important to the response in question. How this is done exactly is outside the scope of this book, since it requires more advanced statistical modeling techniques.

10.4.5 Limitations

Unlike experimental design, which focuses on parameters (factors), the distance approach focuses on Earth models, recognizing that summarizing the complex Earth is difficult with a few parameters and an optimal design is difficult to select when large spatial models are involved. Model selection by distance allows the study of the influence of virtually any "parameter," including parameters that cannot be ranked or given a value. An interesting parameter to study is "model complexity," that is, how complex a model should be built for the purpose at hand: would a simple physical or spatial model suffice or is more detail needed? It has to be realized that this question cannot be solved without actually building some complex models and understanding the interaction of model complexity in combination with other model parameters. It simply means that not all Earth models need to be complex; instead, one could envision building a mixture of simple and complex models and study the sensitivity to the response in question.

In conclusion, the choice of a distance is a burden as well as an opportunity. It allows injecting the purpose of modeling into the question of uncertainty, but a distance needs to be chosen and that distance needs to be related to the decision question at hand. This can be a subjective choice and one may not know well a priori whether or not a suitable distance has been retained. Imagine, for example, that a distance is selected that is completely wrong, that is, it has no relation to the decision question or response function at hand. Model selection would then result in random selection, which is not very efficient, however not necessarily biased. A poorer distance will require selecting more models to retain a model of uncertainty reflective of the uncertainty determined to be present in the modeling of the system in question. Statistical techniques, such as bootstrap sampling, can be used to determine such accuracy but are outside the scope of this book.

Further Reading

Bishop, C.M. (2006) *Pattern Recognition and Machine Learning*, Springer Verlag.

Fisher, R.A. (1935) *The Design of Experiments*, 8th edn, Hafner Press, New York, 1966.

Ryan, T.P. (2007) *Modern Experimental Design*, John Wiley & Sons, Inc.

Scheidt, C. and Caers, J. (2009) Uncertainty quantification in reservoir performance using distances and kernel methods – application to a West-Africa deepwater turbidite reservoir. *SPE Journal*, 118740-PA, Online First, doi: 10.2118/118740-PA.

White, C.D., Willis, B.J., Narayanan, K., and Dutton, S.P. (2001), Identifying and estimating significant geologic parameters with experimental design, SPE 74140. *SPE Journal*, **6**(3), 311–324.

11

Value of Information

There is no intrinsic value in collecting data unless it can influence a specific decision goal. The aim of collecting more data is to reduce uncertainty on those parameters that are influential to the decision making process. Therefore, value of information is dependent on the particular decision problem faced before taking any additional data. As a result, three key components are involved in determining the value of information: (1) prior uncertainty, (2) information content of data and (3) the decision problem.

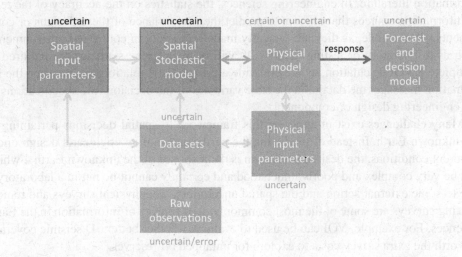

11.1 Introduction

In many practical situations, we are faced with the following question: given the uncertainty about the Earth, do we proceed with making a decision (as explained in Chapter 4) or do we gather more information in order to reduce the uncertainty around the outcomes that influence our decision? In oil reservoirs, examples of gathering information to reduce uncertainty include conducting a seismic study, doing more outcrop studies, coring a well,

---

running a well-test analysis, consulting an expert, running logging surveys, doing a reservoir modeling study, and so on. The intuitive reason for gathering information is straightforward: if the information can reduce uncertainty about future outcomes, decisions can be made that have better chances for a good outcome. However, such information gathering is often costly. Questions that arise include (a) is the expected uncertainty reduction worth its cost, (b) if there are several potential sources of information, which one is most valuable, and (c) which sequence of information sources is optimal. This type of question is framed under "the value of information problem." This question is not trivial to answer because it is necessary to assess this value before any measurement is taken.

Decision analysis and value of information (VOI) have been widely applied to decisions involving engineering designs and tests, such as assessing the risk of failure for buildings in earthquakes, components of the space shuttle and offshore oil platforms. In those fields, gathering information consists in doing more "tests" and if those tests are useful, that is, they reveal design flaws (or lack thereof), then such information may be valuable depending on the decision goal. It seems intuitive that it is necessary to come up with some measure of "usefulness" of the test. Indeed, if the test conducted does not at all inform the decision variable of interest, then there is no point in conducting it. The "degree of usefulness" is termed the "reliability" of the test in the traditional value of information literature. In engineering sciences, the statistics on the accuracy of the tests or information sources that attempt to predict the performance of these designs or components are available, as they are typically made repeatedly in controlled environments, such as a laboratory or testing facility. As will be seen, these statistics are required to complete a VOI calculation, as they provide a probabilistic relationship between the information message (the data) and the state variables of the decision (the specifications of the engineering design or component).

Many challenges exist in applying this framework to spatial decisions pertaining to an unknown Earth. Instead of predicting the performance of an engineer's design under different conditions, the desired prediction is the response of the unknown Earth – which can be very complex and poorly understood and certainly cannot be put in a laboratory – due to some external action and the spatial uncertainty. Geophysical surveys and remote sensing surveys are some of the most commonly used sources of information in the Earth Sciences. For example, VOI can be used to evaluate whether better 3D seismic coverage is worth the extra survey costs to explore for untapped oil reserves.

11.2 The Value of Information Problem

11.2.1 Introduction

A catch-22 situation presents itself when determining VOI of any measurement in the Earth Sciences. VOI is calculated before any information is collected. Note that data collection can be done only once and is rarely repeatable. However, the VOI calculation cannot be completed without a measure that describes how well the proposed measurement resolves what one is trying to predict. This measure is known as the "data reliability measure" and is used to determine the discrimination power of measurements to resolve the Earth characteristics key to the model of uncertainty.

It has been stated that there is no intrinsic value in collecting data unless it can influence a specific decision goal. The aim of collecting more data is to reduce uncertainty on those parameters that are influential to the decision making process. Therefore, VOI is dependent on the particular decision problem faced before taking any additional data. As a result, three key components are involved in determining the VOI:

1 The prior uncertainty of what one is trying to model: the more uncertain one is about some Earth phenomenon, the more the data can possibly contribute to resolving that uncertainty for a particular decision goal.

2 The information content of the data (this will be translated into a data reliability or vice versa): if the data is uninformative it will have no value. But even perfect data (data that resolves all uncertainty) may not help if that doesn't influence the decision question.

3 The particular decision problem: this drives the value assessment on which any VOI calculation is based.

11.2.2 Reliability versus Information Content

Critical to the VOI question is to determine the "reliability" of the data. Suppose we have a data source (e.g., Ground Penetrating Radar, GPR), termed B and would like to resolve some unknown parameter (e.g., the layer thickness), termed A. In general terms, reliability is a conditional probability of the form

$$P(\text{data says} \mid \text{real world is})$$

or in words: "probability of what the data says, given what the real world is." For example (Figure 11.1): it is assumed that the real world or actual groundwater level may have two outcomes, h_1 and h_2, then, if the data source is imperfect, it may sometimes be correct (measuring h_i while the actual value is also h_i) or sometimes wrong (measuring h_j while the actual world is h_i). The table in Figure 11.1 records these frequencies, where B is the random variable representing the data and A is the random variable representing the real world. Note that the data are random variables because it is not known what the data will be in VOI problems.

In Chapter 7, another probability was discussed, named "information content," which is of the type:

$$P(\text{real world is} \mid \text{data says})$$

What is often available is the data reliability, but, as will be seen in the next section, in order to solve the value of information problem, it is necessary to know the information

Reliability $P(B=b\|A=a)$		Actual interval	
		h_1	h_2
The GPR data says	h_1 measured	$P(B=b_1\|A=a_1)$	$P(B=b_1\|A=a_2)$
	h_2 measured	$P(B=b_2\|A=a_1)$	$P(B=b_2\|A=a_2)$
	Total	1	1

Information content $P(A=a\|B=b)$		The GPR data says	
		h_1 measured	h_2 measured
Actual interval	h_1	$P(A=a_1\|B=b_1)$	$P(A=a_1\|B=b_2)$
	h_2	$P(A=a_2\|B=b_1)$	$P(A=a_2\|B=b_2)$
	Total	1	1

Figure 11.1   Reliability frequencies and information content frequencies in a table.

content. Fortunately, Bayes' rule can be used (Chapter 2) to deduce one from the other as follows

$$P(A = a_1|B = b_1) = \frac{P(B = b_1|A = a_1)\,P(A = a_1)}{P(B = b_1)} \tag{11.1}$$

with

$$P(B = b_1) = P(B = b_1|A = a_1)\,P(A = a_1) + P(B = b_1|A = a_2)\,P(A = a_2) \tag{11.2}$$

11.2.3 Summary of the VOI Methodology

The main procedural steps in carrying out a value of information study are outlined below. These steps will be illustrated and explained by application to a simple decision problem:

1 Calculate the expected value of the decision to be made without the information. This is usually carried out using the methodology described in Chapter 4 and depicted using a decision tree. Term this $V_{\text{without data}}$.

2 Formulate the structure of the decision situation to include the new information. Again, this is usually carried out and depicted by adding a new branch to the decision tree.

3 Calculate the expected value of the new branch, $V_{\text{with perfect information}}$, that is, the value *with* perfect information. This can done either by entering "1" and "0" for reliability probabilities in step 5 below. Then the value *of* perfect information is:

$$\text{VOI}_{\text{perfect}} = V_{\text{with perfect information}} - V_{\text{without data}}$$

4 If $VOI_{perfect}$ is negligible or less than the cost of acquiring the information, resort to the decision made in step 1.

5 If $VOI_{perfect}$ is significant, calculate the value when the data are considered to be imperfect information, which has the following substeps:

a Determine the reliability probabilities – P(data says | real world is).

b Calculate the information content (IC) probabilities.

c Enter the IC probabilities in the decision tree and solve the new branch to obtain the $V_{with\ imperfect\ information}$.

d Calculate $VOI_{imperfect}$ as:

$$VOI_{imperfect} = V_{with\ imperfect\ information} - V_{without\ data}$$

$VOI_{imperfect}$ can then be compared with the case of gathering the information or data.

11.2.3.1 Steps 1 and 2: VOI Decision Tree

Suppose we have a simple decision to either do something or not (e.g., clean up or not). Denote this binary decision as having two alternatives "clean" or "do not clean." Now suppose an uncertain event A that has two possible outcomes, a_1 and a_2, namely a_1 = "soil is sandy" and a_2 = "soil is clayey;" in case a_1 spreading of contaminants is less likely to occur.

If we take action "clean up," then the cost is fixed, namely CleanCost (a negative value or payoff of '$- C$'). If we do not clean up, then we could be lucky, that is, the soil contains more clay and contaminants do not spread and we do not need to pay anything. If, however, the soil is more sandy then pollution can spread and we have an extra penalty; in that case the cost is: CleanCost + Penalty or $-(C + P)$. This defines our initial decision problem, which can be represented by the tree in Figure 11.2. The tree can be solved revealing which action has the lowest cost. Value (or cost) of the optimal action is denoted as $V_{without\ data}$.

Next consider gathering more data. In terms of decision analysis, this is simply another alternative, next to the actions "to clean up" and "not to clean up." We now have a third alternative in a new decision tree (Figure 11.3). Consider now the data source, denoted as B, which has two possible outcomes, b_1 = "data indicates sandy soil" and b_2 = "data indicates clayey soil." Once the data are gathered, we resort to our initial decision problem and attach this to each of the branches $B = b_1$ and $B = b_2$. What has changed in these branches is that the probabilities associated with the uncertain event A are now conditional, namely dependent on the outcome of the data. If the data are perfect, then the data will fully resolve the uncertain event, that is, they will reveal whether or not we have

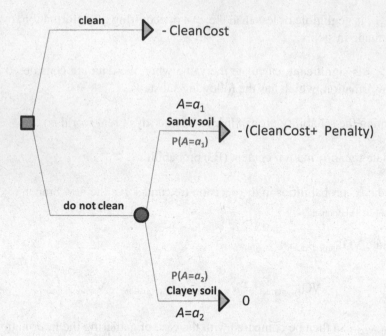

Figure 11.2   Initial decision tree.

sandy soil. If, however, the data are imperfect, which means that even when the data says "sandy soil" the actual soil may be clay, then it is necessary to determine the information content probabilities $P(A = a_i \mid B = b_j)$, $i,j = 1$ to 2. Recall that these probabilities are deduced from the reliabilities (Equation 11.1), which should be made available to the decision maker. Remaining is the specification of the marginal probabilities $P(B = b_1)$ and $P(B = b_2)$, which are given by Equation 11.2. Once all probabilities are known, the tree can be solved, and a value (or cost) for the third branch can be calculated. Denote this cost as $V_{\text{with data}}$. The value of information is then defined as:

$$\text{VOI} = V_{\text{with data}} - V_{\text{without data}}$$

Note that the decision tree does not contain the cost of taking the survey. In decision analysis such cost is termed a "sunk cost," that is, if you decide to gather data then you will always incur this cost.

11.2.3.2 Steps 3 and 4: Value of Perfect Information

Consider for illustration purposes some actual values for the decision problem of Figure 11.2 and 11.3. The decision tree is completed by entering the prior probabilities, which are assumed to be known:

$$P(A = a_1) = 0.4 \text{ and } P(A = a_2) = 0.6$$

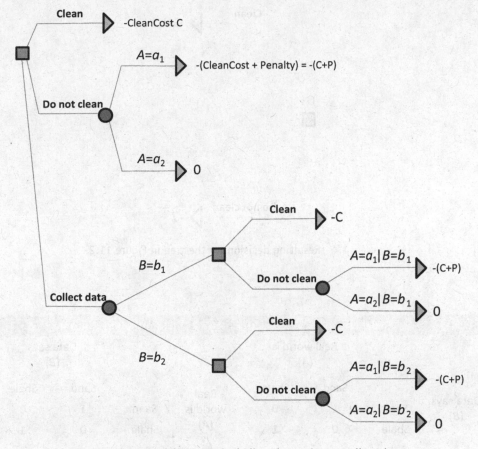

Figure 11.3   Decision tree, including the option to collect data.

The costs are known, too, and assumed deterministic:

CleanCost = 10; Penalty = 7.5; (in units of 10 000 dollars for example)

Solving the tree according to the techniques in Chapter 4 yields the result in Figure 11.4, meaning that the best decision is to not clean up, or:

$$V_{\text{without data}} = -7$$

Returning to the branch involving collecting more data. With perfect information we have that (Figure 11.5):

$$P(B = b_1|A = a_1) = P(A = a_1|B = b_1) = 1 \text{ and } P(B = b_2|A = a_1)$$
$$= P(A = a_1|B = b_2) = 0$$
$$P(B = b_1) = P(A = a_1) \text{ and } P(B = b_2) = P(A = a_2)$$

These values are inserted in the tree of Figure 11.3 resulting in Figure 11.6.

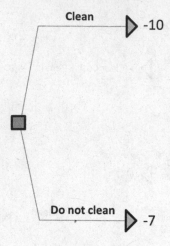

Figure 11.4   Resulting decision for the tree in Figure 11.2.

Reliability				Information content			
		Real world is (A)				Data says (B)	
		Sand	Shale			Sand	Shale
Data says (B)	Sand	1	0	Real world is (A)	Sand	1	0
	Shale	0	1		Shale	0	1

Figure 11.5   Reliabilities and information content with perfect information.

Figure 11.6   Decision tree with perfect information.

Given this result, the value with perfect information is:

$$\text{VOI}_{\text{perfect}} = V_{\text{with perfect information}} - V_{\text{without data}} = -4 - (-7) = 3$$

The value of perfect information seems reasonably high ($30 000) which means that it is meaningful to calculate the value of imperfect information if the cost of the survey is less than this number. The value of imperfect information is dependent on the reliabilities (either directly given or calculated from the information content probabilities) and is explained next.

11.2.3.3 Step 5: Value of Imperfect Information

Consider the reliability probabilities (given) and the information content probabilities (calculated from the reliabilities) in Figure 11.7 chosen for this example. Note the asymmetry in the data reliability of "shale" vs presence of "sand." From this it can be calculated that:

$$P(B = b_1) = P(B = b_1|A = a_1)P(A = a_1) + P(B = b_1|A = a_2)P(A = a_2) = 0.37$$
$$P(B = b_2) = P(B = b_2|A = a_1)P(A = a_1) + P(B = b_2|A = a_2)P(A = a_2) = 0.63$$

Inserting all probabilities in red in Figure 11.8 gives the solution

$$\text{VOI}_{\text{imperfect}} = V_{\text{with imperfect information}} - V_{\text{without data}} = -5.8 - (-7) = 1.2$$

meaning the data should cost only $12000 to be valuable. An interesting question that can be considered is: what should the reliability probabilities be such that the information is no longer valuable, that is, $\text{VOI}_{\text{imperfect}} = 0$; if we only consider changing the reliability of sand (currently 0.7/0.3) we find that the minimum reliability for the information to have value is $P(B = b_1 | A = a_1) = 0.30$. This can then be used by modelers to decide which information to gather or which measurement device to deploy, or how accurate the measurement should be taken, if of course the reliability probabilities for these data sources are available.

Reliability		Real world is (A)		Information content			Data says (B)	
		Sand	Shale				Sand	Shale
Data says (B)	Sand	0.70	0.15	Real world is (A)		Sand	0.75	0.19
	Shale	0.30	0.85			Shale	0.25	0.81

Figure 11.7   Reliability probabilities and information content of imperfect data.

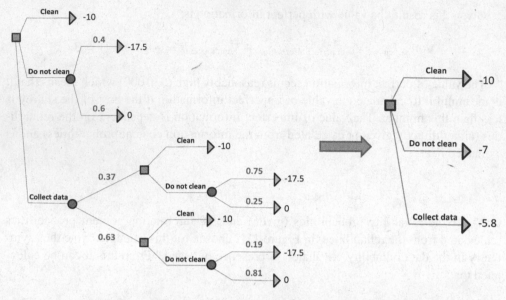

Figure 11.8   Final decision tree for calculating the value of imperfect information.

11.2.4 Value of Information for Earth Modeling Problems

11.2.5 Earth Models

In the above a VOI calculation was presented for a relative simple setting: we have some unknown A, which can have a few outcomes, and some data source B, also with a few outcomes, from which conditional probabilities were deduced. What if the unknown is that part of the Earth targeted for modeling using the techniques in Chapters 5–8?

The most general way of representing uncertainty is to generate a set of Earth models that includes all sources of uncertainty prior to collecting the data in question for the value of information assessment. Such a large set of Earth models may be created using a variety of techniques, including the variation of subsurface structures, that is, faults and/or layering, uncertainty in the depositional system governing important properties (permeability, porosity, ore grade, saturation, etc.) contained within these structures as well as the spatial variation of such properties simulated with geostatistical techniques.

In terms of spatial modeling two components of uncertainty modeling in creating such Earth models were discussed (Figure 11.9). Firstly, the prior uncertainty on input parameters needs to determined. These input parameters could be as simple as the range or azimuth of a variogram model that is uncertain and is described by a probability density function (pdf) or it could be a set of training images, each with a prior probability of occurrence. Next, a particular outcome of the spatial input parameter(s) is (are) drawn and a stochastic simulation algorithm generates one or several Earth models for that given set of parameters. The two components of uncertainty are: uncertainty of the input parameters and uncertainty due to spatial variation for a given set of input parameters.

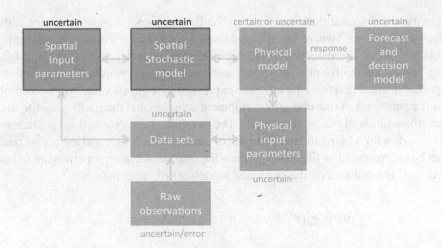

Figure 11.9 Input uncertainty and spatial uncertainty.

In presenting the value of information calculation, it will be assumed, for ease of presentation and notation that there is only one spatial input parameter, denoted by the random variable Θ. For convenience, it will also be assumed that this variable is discrete and has two possible outcomes θ_1 and θ_2. Some prior probability is stated on θ_1 and θ_2. θ_1 and θ_2 could denote two possible variogram models, two alternative training images, or two structural model concepts. Once such outcome θ is known, then several alternative Earth models can be created. Such Earth models will be denoted

$$\mathbf{z}^{(t)}(\theta) \quad t = 1, \ldots, T$$

where T is the total number of such Earth models. Evidently they depend on θ. Note that $\mathbf{z}$ is multivariate since many correlated (or uncorrelated) spatial variables can be generated (porosity, permeability, soil type, concentrations, fault position and so on) in a single Earth model. Further input uncertainty will be considered as the most sensitive factor in the decision problem and that spatial uncertainty is less consequential, hence the goal is to reduce uncertainty on θ in order to hopefully make a more informed decision.

11.2.6 Value of Information Calculation

Similar to the simple case, the determination of the value of information is described in three parts. Firstly, there is no value in data unless it can influence the spatial decision considered (i.e., choosing between several different and possible locations for a well or mine extraction schemes). Therefore, the outcome of a spatial decision will be expressed in terms of value as related to a specific decision outcome. A general description of the value of information problem for spatial models will be provided before proceeding to an actual case study.

Many types of spatial decisions exist in the Earth Sciences. In oil recovery, different development schemes (where to drill wells and what kind of wells) represent different possible actions or alternatives a to the decision of how to develop a particular field. In mining, several alternative mine plans represent such actions or several clean up strategies for a polluted site. Therefore, the outcome expressed in terms of value will be a combination of the action taken (the chosen development scheme) and the Earth response to these actions (the amount of oil/ore recovered). The possible alternative actions are indexed by $a = 1, \ldots, A$, with A being the total number of alternative actions, and the action taken on the Earth is represented as function g_a. Since the true subsurface properties are unknown, the action g_a is simulated on the generated models $\mathbf{z}^{(t)}(\theta)$ such that:

$$v_a^{(t)}(\theta) = g_a(\mathbf{z}^{(t)}(\theta)) \qquad a = 1, \ldots, A \quad t = 1, \ldots, T.$$

Note that the value $v_a^{(t)}$ is a scalar. As discussed in Chapter 4, value can be expressed in a variety of terms; however, monetary units (usually expressed in net present value, NPV) are conceptually the most straightforward.

For any situation, the decision alternative that results in the best possible outcome should be chosen. However, this is difficult to determine in advance because of uncertainty regarding how the Earth will react to any proposed action. The values in the above equation could vary substantially due to such uncertainty. Based on that variation we determine the value without data $V_{\text{without data}}$ in the case where Θ has two categories θ_1 and θ_2 as follows:

$$V_{\text{without data}} = \max_a \left(\sum_{i=1}^{2} P(\Theta = \theta_i) \frac{1}{T_{\theta_i}} \sum_{t=1}^{T_{\theta_i}} v_a^{(t)}(\theta_i) \right) \qquad a = 1, \ldots, A$$

Analyzing this equation a bit more:

- $\frac{1}{T_{\theta_i}} \sum_{t=1}^{T_{\theta_i}} v_a^{(t)}(\theta_i)$ is the average value (in \$) for that action (e.g., clean up) and for that value of θ (e.g., fluvial depositional system exists).

- $\sum_{i=1}^{2} P(\Theta = \theta_i) \frac{1}{T_{\theta_i}} \sum_{t=1}^{T_{\theta_i}} v_a^{(t)}(\theta_i)$ is the expected value of that average because the prior probabilities for Θ may not be equal.

- $\max_a$: takes the maximum over all the actions of this expected value.

$P(\Theta = \theta_i)$ represents the prior uncertainty for geologic input parameter θ_i and T_θ is the number of Earth models generated when $\Theta = \theta_i$. The computation of V_{prior} to N_θ

categories can be generalized as follows:

$$V_{\text{without data}} = \max_a \left(\sum_{i=1}^{N_\theta} P(\Theta = \theta_i) \frac{1}{T_{\theta_i}} \sum_{t=1}^{T_{\theta_i}} v_a^{(t)}(\theta_i) \right) \qquad a = 1, \ldots, A.$$

Turn now to the value with data ($V_{\text{with data}}$), which needs to involve the reliability probabilities. The first problem is that the data do not directly inform Θ. For example, a geophysical measurement may provide seismic readings or electromagnetic readings; it does not directly provide a reading of θ, such as for example: what is the depositional system? Often such raw measurements need to be further processed and interpreted (as was shown in Figure 11.10). The second problem is that no data have been taken yet, so there is nothing to process or interpret. To get to the reliability probabilities the scheme in Figure 11.10 is considered.

In summary, the following steps are followed to obtain the probability reliabilities for the binary case with two possible outcomes θ_1 and θ_2:

1 Create an Earth model using parameter θ_j with Θ drawn from prior $P(\Theta = \theta_i)$ $i = 1,2$.

2 Apply the forward model that models the physical relationship between the data and the Earth model.

3 Term the response obtained from this forward model the simulated data $\mathbf{d}$.

4 Obtain an interpretation of Θ from $\mathbf{d}$; term this interpretation θ_j^{int}.

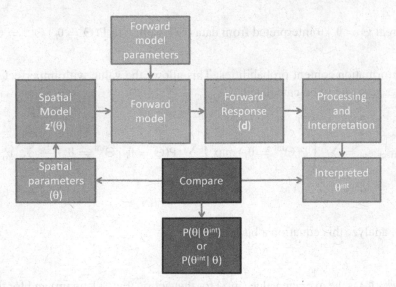

Figure 11.10 Overview scheme for obtaining reliability probabilities for spatial value of information problems.

5 Compare the interpretation θ_j^{int} with the θ_j in step 1.

6 If they match, then call this "Success for θ_j"

7 Redo steps 1–6 for a sufficient number of times, say N.

8 Calculate the reliability probabilities as follows:

$$P(\Theta^{\text{int}} = \theta_1 | \Theta = \theta_1) = \frac{\text{\# of succes for } \theta_1}{N} \Rightarrow P(\Theta^{\text{int}} = \theta_2 | \Theta = \theta_1)$$

$$= 1 - \frac{\text{\# of succes for } \theta_1}{N}$$

$$P(\Theta^{\text{int}} = \theta_2 | \Theta = \theta_2) = \frac{\text{\# of succes for } \theta_2}{N} \Rightarrow P(\Theta^{\text{int}} = \theta_1 | \Theta = \theta_2)$$

$$= 1 - \frac{\text{\# of succes for } \theta_2}{N}$$

Much of this scheme is similar to the scheme for solving inverse problems except that: (1) the forward model now models the physics of the data gathering tool and its response when applied to the Earth (the Earth being modeled by an Earth model); (2) it is not an inverse problem *per se* because there are no data that can be used for inverse modeling. The steps above need to be repeated many times (N times in the above summary), that is, for each new Earth model a new interpretation is generated, then, based on the frequencies obtained of correct and incorrect, the reliability or information content probabilities can be deduced, denoted now as

$$P(\text{real } \Theta = \theta_i \,|\, \theta \text{ interpreted from data } (\Theta^{\text{int}} = \theta_j)) \text{ or } P(\Theta = \theta_i \,|\, \Theta^{\text{int}} = \theta_j)$$

for the information content probabilities. This allows the value with imperfect information $V_{\text{with imperfect data}}$ to be calculated:

$$V_{\text{with imperfect data}} = \sum_{j=1}^{N_\theta} \left(P(\Theta^{\text{int}} = \theta_j) \max_a \left(\sum_{i=1}^{N_\theta} P(\Theta = \theta_i \,|\, \Theta^{\text{int}} = \theta_j) \frac{1}{T_{\theta_i}} \sum_{t=1}^{T_{\theta_i}} v_a^{(t)}(\theta_i) \right) \right)$$

$$a = 1, \ldots, A$$

Again, analyze this equation a bit more:

- $\frac{1}{T_{\theta_i}} \sum_{t=1}^{T_{\theta_i}} v_a^{(t)}(\theta_i)$ is the average value (in \$) for that action (e.g., clean up) and for that value of θ (e.g., a fluvial depositional system exists).

- $\sum_{i=1}^{N_\theta} \mathrm{P}(\Theta = \theta_i \mid \Theta^{\mathrm{int}} = \theta_j) \frac{1}{T_{\theta_i}} \sum_{t=1}^{T_{\theta_i}} v_a^{(t)}(\theta_i)$ is then the expected value of this average for all possible θ that have been interpreted from the data.

- $\max_a$: takes the maximum over all the actions of this expected value.

Figure 11.11 provides an example decision tree for the case when θ is a binary variable.

In the case of perfect information, it is known in advance that the data will always reveal the real θ, so there is no need for applying the workflow of Figure 11.11, and the value with perfect information becomes

$$V_{\text{with perfect information}} = \sum_{i=1}^{2} \mathrm{P}(\Theta = \theta_i) \max_a \left(\frac{1}{T_{\theta_i}} \sum_{t=1}^{T_{\theta_i}} v_a^{(t)}(\theta_i) \right) \quad a = 1, \dots, A.$$

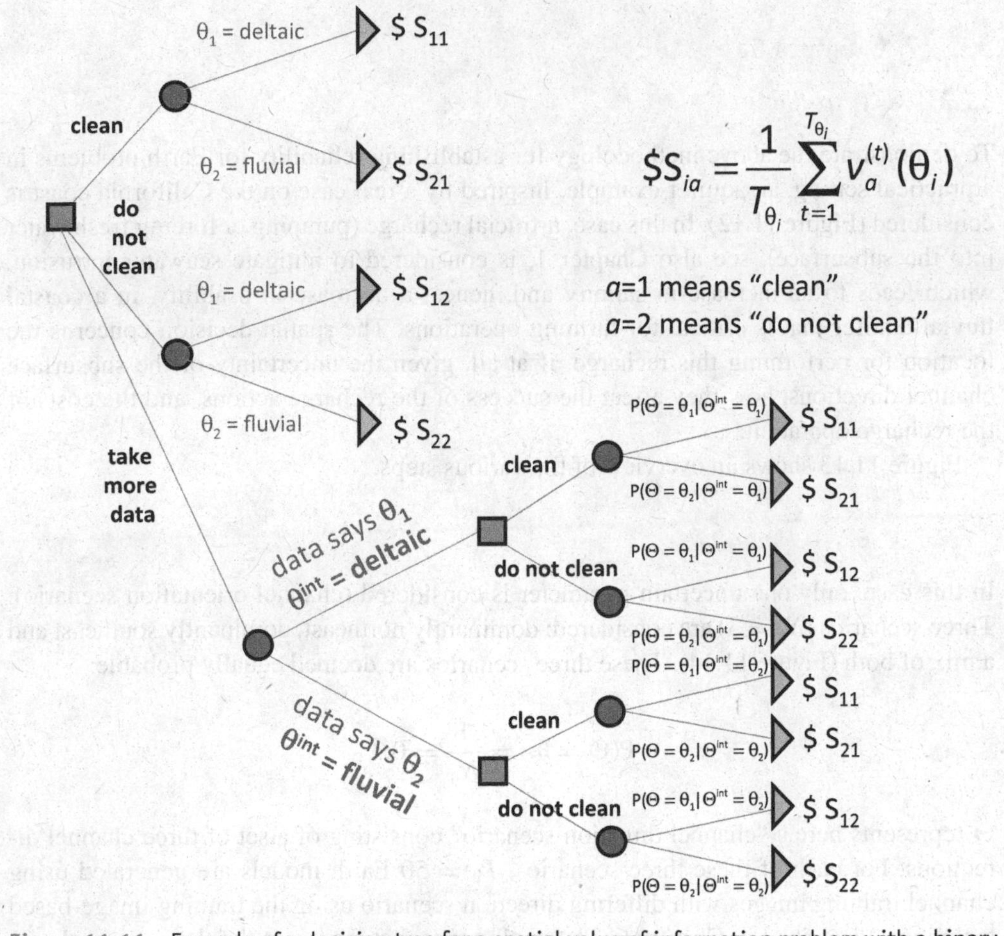

$$\$S_{ia} = \frac{1}{T_{\theta_i}} \sum_{t=1}^{T_{\theta_i}} v_a^{(t)}(\theta_i)$$

$a=1$ means "clean"
$a=2$ means "do not clean"

Figure 11.11 Example of a decision tree for a spatial value of information problem with a binary decision and a binary variable.

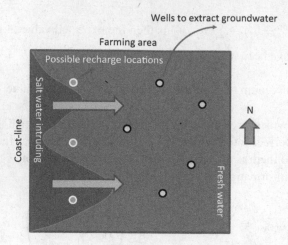

Figure 11.12   Outline of the salt water intrusion problem.

11.2.7 Example Case Study

11.2.7.1 Introduction

To demonstrate the above methodology for establishing reliability for Earth problems in a practical setting, an aquifer example, inspired by a real case on the California coast is considered (Figure 11.12). In this case, artificial recharge (pumping or forcing fresh water into the subsurface), see also Chapter 1, is considered to mitigate seawater intrusion, which leads to an increase in salinity and, hence, a decrease in usability, in a coastal fluvial aquifer that is critical for farming operations. The spatial decision concerns the location for performing this recharge, if at all, given the uncertainty of the subsurface channel directions, how they affect the success of the recharge actions, and the costs of the recharge operation.

Figure 11.13 shows an overview of the various steps.

11.2.7.2 Earth Modeling

In this case, only one uncertain parameter is considered (channel orientation scenario). Three scenarios ($N_\theta = 3$) are considered: dominantly northeast, dominantly southeast and a mix of both (Figure 11.14). These three scenarios are deemed equally probable:

$$P(\Theta = \theta_i) = \frac{1}{N_\theta} = 0.3$$

Θ represents here a "channel direction scenario" consisting of a set of three channel directions. For each of these three scenarios, $T_\theta = 50$ Earth models are generated using channel training images with differing direction scenario using the training image-based techniques described in Chapter 6. The angle maps that represent the three channel scenarios and example 2D Earth models generated from each are illustrated in Figure 11.14.

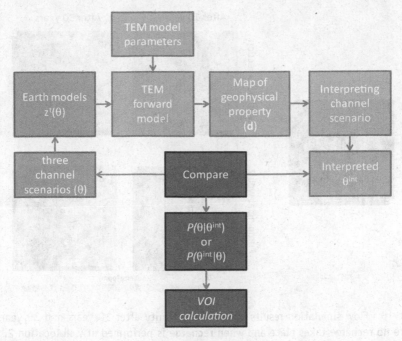

Figure 11.13 Various steps needed to solve the VOI problem for the case study of Figure 11.12.

11.2.7.3 Decision Problem

Four alternatives to the recharge decision are identified as (1) no recharge, (2) a central recharge location, (3) a northern recharge location, and (4) a southern recharge location (Figure 11.12). Flow simulation is performed for all combinations to assess the response to each such action. The flow simulation outputs the evolution of salt water intrusion over

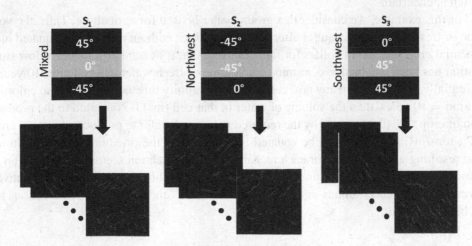

Figure 11.14 Schematic showing how various Earth models are created by changing locally the channel orientation in three different scenarios.

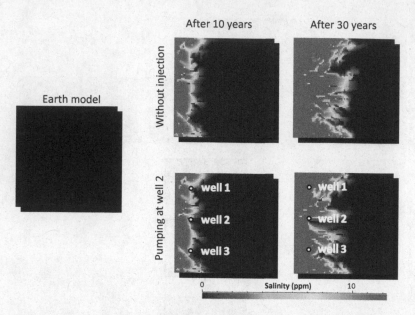

Figure 11.15 Flow simulation results in terms of salinity after 10 years and 30 years for the case where no recharge takes place and when recharge is performed at well location 2.

time due to the combined extraction of ground water for farming operations as well as the artificial recharge. Given that there are four recharge options and three channel direction scenarios (specifically the 50 lithology Earth models within each) this results in a total of $50 \times 3 \times 4 = 600$ flow simulations. An example of a flow simulation applied to one of the Earth models is shown in Figure 11.15, demonstrating the effect of activating pumping operations in well 2. The well does prohibit somewhat the intrusion of salt water into the aquifer, but such intrusion is also clearly dependent on the subsurface heterogeneity, which is uncertain.

For this example, we consider that groundwater is used for agriculture. Thus, the volume of fresh water in the aquifer after 30 years, with or without recharge, is equated into potential crop revenue in dollars for farmers. Note that for each Earth model a flow simulation has been obtained to determine the volume of fresh water in a cell after 10 years. Essentially, if the water in any grid cell is below a salinity threshold of 150 ppm chloride at time $= 10$ years, then the volume of water in that cell (m^3) is converted to the production of crop "X" (this is given by the required volume of water to produce one ton of crop "X": tons/m^3). This can then be equated into dollars by the price of crop "X" ($/ton). The resulting average value for each recharge channel direction scenario combination is shown in Figure 11.16. The prior value of these possibilities and decision alternatives given the prior uncertainties is $12.47 million with as the chosen decision action: no recharge (needed).

11.2.7.4 The Possible Data Sources

Transient or time-domain electromagnetic (TEM) data are considered to help determine aquifer heterogeneity, specifically the channel orientation. TEM works with a transmitter

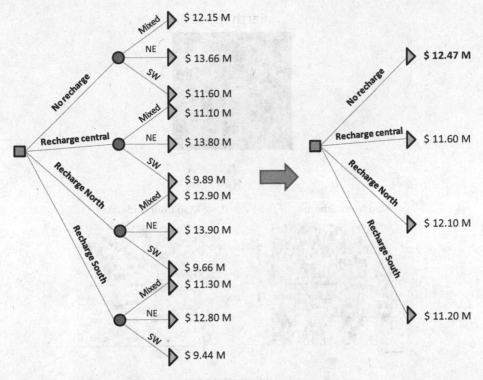

Figure 11.16 Decision tree and solution.

loop that turns on and off a direct current to induce currents and magnetic fields into the subsurface; meanwhile, a larger receiver loop measures the changing magnetic field response of induced currents in the subsurface. Two types of TEM survey can be carried out: a land-based survey, where the equipment is deployed on the ground, and an airborne survey, requiring a plane to fly over the area of interest.

The data (the magnetic field response in time) must then be inverted to a layered model of electrical resistivity and thickness values, which is achieved through geophysical inversion (not shown here). The recovered electrical resistivity can be an indication of the lithology type as clay (non-sand) typically has an electrical resistivity less than 30 ohm-m, whereas sand is usually greater than 80 ohm-m. Therefore, from the TEM measurement, a map indicating lithology can be obtained, but since TEM is only an indirect measure of lithology, such a map does not provide a correct image of the subsurface lithology, meaning that some form of reliability needs to be deduced. Since no data are yet available, geophysical forward modeling and processing need to be applied to all 150 Earth models to obtain 150 lithology-indicating maps. Figure 11.17 shows such a lithology indicating map obtained in case of the airborne and land-based TEM for one Earth model. Clearly, the airborne survey provides greater detail than the land-based survey.

11.2.7.5 Data Interpretation

It is necessary now to compare the channel orientation of each Earth model (θ) with the channel orientation of the lithology maps obtained from the geophysical survey applied

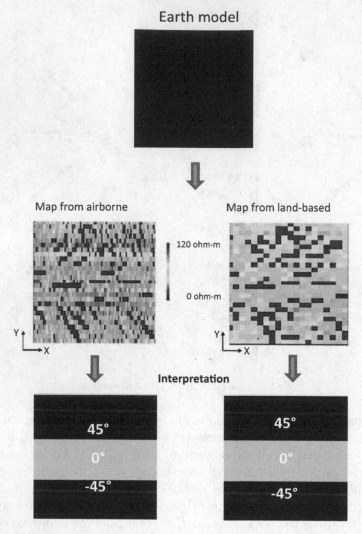

Figure 11.17 Obtaining lithology-indicating maps from an Earth model and data interpretation through image processing.

to each Earth model. A simple image processing tool (not discussed here) turns the geophysical image into a map of interpreted channel scenario. This interpretation, θ^{int}, is then compared with the actual θ. If such comparison is done for the 150 Earth models, the frequencies listed in Figure 11.18 are obtained.

The two reliabilities of airborne and land-based measurements are quite different. This is due to the difference in resolution with which each technique takes measurement. The resolution of channel orientation is decreased for land TEM with fewer locations of measurements. Utilizing the reliability measures of Figure 11.18, the VOI$_{imperfect}$ for airborne and land-based TEM are ~$300 000 and $0. Compared to VOI$_{perfect}$ ($350 000) the influ-

Reliability Airborne TEM				
		Real world is (A)		
		Mixed	NE	SW
Data says (B)	Mixed	0.96	0	0
	NE	0	0.98	0
	SW	0.04	0.02	1

Reliability land-based TEM				
		Real world is (A)		
		Mixed	NE	SW
Data says (B)	Mixed	0.64	0.24	0.37
	NE	0.04	0.49	0.26
	SW	0.32	0.27	0.37

Figure 11.18 Reliabilities for land-based and airborne TEM in discriminating channel scenario.

ence of the reliability measure is apparent. If the price of acquiring airborne TEM is less than $300 000, then it is deemed a sound decision to purchase this information.

Further Reading

Bratvold, R.B., Bickel, J.E., and Lohne, H.P. (2009) Value of information in the oil and gas industry: past, present, and future. *SPE Reservoir Evaluation & Engineering*, **12**(4), 630–638.

Howard, R.A. (1966) Information value theory. *IEEE Transactions on Systems Science and Cybernetics* (SSC-2), 22–26.

12

Example Case Study

12.1 Introduction

12.1.1 General Description

In this chapter the example case study introduced in Chapter 4 is revisited. All elements related to modeling uncertainty and decision analysis to solve this problem have now been discussed. The example case study is constructed (synthetic). The data and programs needed to solve this problem are available at http://uncertaintyES.stanford.edu. The idea here is to provide you with data and code that you can run and test yourself. The example discussed here is a simplification from an actual case but can easily be extended to include more complexity within the given software. The software platform used is S-GEMS (the Stanford Geostatistical Earth Modeling software, which is freely available).

Here again is the problem description:

Consider an area, as shown in Figure 4.1, where a point source of pollution due to leakage of chemicals was discovered in the subsurface. This pollution source is close to a well that is used to supply drinking water. While it hasn't happened yet, some speculate that due to the geological nature of the subsurface, this pollution may travel to the drinking well. The study area consists of porous unconsolidated sand in a non-porous clay material. The deposit lies over an impermeable igneous rock. Some basic geological studies have revealed that this is an alluvial deposit. Some geologists argue that the main feature of these deposits is channel ribbons, such as in Figure 4.1. Only very basic information is known about this type of channel based on analog information and an outcrop in a nearby area, but most important is the channel orientation Θ_{ch}, which may impact the direction of the contaminant flow, in fact two possibilities are assessed:

$$P(\Theta_{ch} = 150°) = 0.4 \quad P(\Theta_{ch} = 50°) = 0.6$$

However, geologists of equal standing claim that this area does not contain sand channels but sand bars (half-ellipsoid) that tend to be smaller than long sinuous sand channels.

Modeling Uncertainty in the Earth Sciences, First Edition. Jef Caers.
© 2011 John Wiley & Sons, Ltd. Published 2011 by John Wiley & Sons, Ltd.

Similarly, the orientation of these elliptical bars is important and two possibilities are assessed:

$$P(\Theta_{bar} = 150°) = 0.4 \quad P(\Theta_{bar} = 50°) = 0.6$$

Some believe that there are enough barriers in the subsurface to prevent such pollution and claim that the pollution will remain isolated and that clean up would not be required. Some believe that even if the pollutant reaches the drinking well the concentration levels will be so low because of mixing and dilution in the aquifer that it will not pose a health concern and hence investment in clean up is not required.

The local government has to make a decision in this case, which is to act and start a clean up operation (which is costly for tax payers) or do nothing, thereby avoiding the clean up cost but potentially be sued later in court (by local residents) for negligence when drinking water is actually contaminated. What decision would the local government make? Clean up or not? How would they reach such decision?

In addition, information on geological object dimensions is available in Figure 12.1 in units of grid cells.

Four 3D training images are constructed, each one reflective of the geological scenario (two possibilities) and the unknown channel orientation (two possibilities) (Figure 12.2).

In addition to solving this decision problem, the value of information problem will also be considered. Considering two types of information available, a geophysical data source and a geological data source, the following are known:

1 A geophysical survey that will reveal with given reliability the paleo-direction of deposition (θ). Based on past experience with this type of geophysical survey, the orientation of either bars or ribbons is known to be revealed with the reliability probabilities shown in Figure 12.3.

2 A detailed geological study that will reveal with certainty the depositional model S (full reliability or perfect information), either S = channel scenario or S = bar scenario.

	Channel scenario	Bar scenario
Width	6	6
Thickness	4	6
Length	40	40
Proportion	30%	30%
Position	random	random
Wavelength	25	N/A
Amplitude	3	N/A

Figure 12.1 Specification of known object parameters for each scenario.

Figure 12.2   3D training images with channels and bars at 50 and 150 degrees.

In summary, the following questions need to be answered:

1 Do we clean up or not?

2 What are the most important factors influencing contaminant transport?

3 Do we buy the additional geophysical data or not?

4 Do we buy the additional geological data or not?

Before addressing these questions, a little more information is needed on costs and how contaminant transport is simulated.

Reliability				Information content			
		actual orientation				orientation from geophysics	
		150	50			150	50
orientation from geophysics	150	0.80	0.20	actual orientation	150	0.73	0.14
	50	0.20	0.80		50	0.27	0.86

Figure 12.3   Reliability probabilities and information content for the geophysical data.

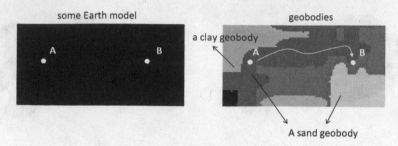

Figure 12.4   Definition of a geobody.

12.1.2 Contaminant Transport

In a detailed study a contaminant transport model would be run to predict for a given Earth model whether or not the contaminant has reached the drinking well. In this study, it will be assumed that the contaminant will travel to the drinking well if, and only if, a geological connection exists between the point source and the drinking well. It will be assumed that the clay is impermeable; hence, if the drinking well and point source lie in the same sand body then the drinking well will get contaminated, otherwise no contamination is going to take place. In order then to determine whether or not there is a connection we calculate, for a given Earth model, the geobodies of that Earth model. Figure 12.4 gives an example. When the Earth model consists of a binary system (just two categories present, such as sand/shale), then a geobody consists of all cells belonging to the same volume. In Figure 12.4, the locations A and B lie in the same "sand" geobody; hence, if a contaminant is released from point A, then it is assumed, in this case study, that it will eventually travel to point B.

12.1.3 Costs Involved

There are several costs related to the problem at hand, these cost are assumed to be deterministic (which is doubtful for a lawsuit). The most relevant of these are the following:

Cost of cleaning up the contamination, $	1 500 000
Cost of lawsuit in case of contamination, $	5 000 000
Cost of geological study, $	50 000
Cost of geophysical study, $:	100 000

12.2 Solution

12.2.1 Solving the Decision Problem

To solve the decision problem we refer to the decision tree in Chapter 4 (Figure 4.15). Needed for the decision tree are the frequencies of "connection" in both channel/sand bar scenario for each angle (50 and 150°), as a "connection" leads to contamination. To calculate these frequencies, the following steps are taken:

1 Consider a given geological scenario s and a given direction θ.

2 Build N Earth models, using 3D training-image based modeling techniques discussed in Chapter 6 with as 3D training image the one belonging to the given geological scenario S and direction θ.

3 Calculate the sand geobodies for each Earth model.

4 Count how many times the drinking well and contaminant source lie in the same sand geobody; denote that count as n.

5 The probability of contamination (connection) is then n/N.

6 Repeat this for all geological scenarios and directions.

Using these simulated Earth models the table of probabilities in Figure 12.5 is obtained.

These frequencies then enter the tree of Figure 4.15, which is solved in Figure 4.16. The decision is to clean up because the expected value of not cleaning up is clearly larger.

12.2.2 Buying More Data

12.2.2.1 Buying Geological Information

Consider now the value of information problem. Firstly, build a new decision tree that reflects the alternative action "buy more data." Consider each data source separately, that is, two alternative actions present itself: "buy geophysical data" and "buy geological data."

Consider first the tree that involves only buying geological data. The new branch in Figure 12.6 reflects this action. If new geological data are purchased, then this new geological data will either reveal the channel scenario or the bar scenario and it is known that

Scenario	Probability of contamination
Channels – 50°	55%
Channels – 150°	89%
Bars – 50°	2%
Bars – 150°	41%

Figure 12.5 Probabilities of contamination for each scenario.

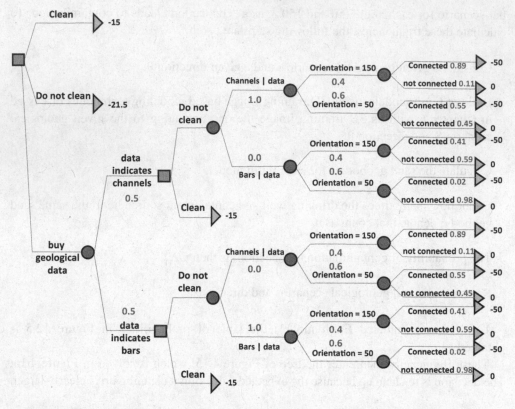

Figure 12.6   Decision tree for the value of information problem "buying geological data."

this will be revealed with certainty (that is if the information is bought). In the tree this is reflected by probability values of one or zero for the conditional events "channels|data" and "bars|data." We can decide whether to clean up or not, and the expected value of these two alternatives will depend on the orientation, which remains uncertain, and the probability of the contaminants traveling to the drinking well, which depends on the geological scenario revealed to us and the orientation. Once the tree is established it is necessary to assign to each uncertainty node or branch a probability. We know there is a 50/50 chance of each scenario occurring and know that the geological data will reveal this with certainty, so the "interpreted scenario from data" is the same as the "true scenario present" and we know that occurs with a 50/50 chance. The probability associated with the two possible orientations is also known and that this variable is independent of the geological scenario. The decision tree is now complete and can be solved. The solution is given in Figure 12.7. The value of information is then

$$\text{VOI}_{\text{geo}} = -11.90 - (-15) = 3.10$$

which is more than the cost of the geological study, which is 0.5.

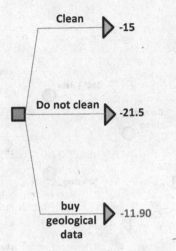

Figure 12.7 Solution of the value of (geological) information problem.

12.2.2.2 Buying Geophysical Information

Geophysical information is not perfect because of the reliabilities in Figure 12.2. Consider first the hypothetical case where the geophysical data are considered perfect, hence all reliability probabilities are either 1 or 0, see the decision tree in Figure 12.8. If perfect geophysical information has no value, then clearly imperfect geophysical information has none either and buying such data should not be considered under the given conditions. From the tree in Figure 12.9 we deduce that

$$\text{VOPI}_{geoph} = -12.96 - (-15) = 2.04$$

which is more than the cost of geophysical data, namely, 1.0. However, the value of imperfect information is

$$\text{VOI}_{geoph} = -14.58 - (-15) = 0.42$$

which is less than the cost of that data. Under the given circumstances, it seems best to buy geological data instead of geophysical data.

12.3 Sensitivity Analysis

In Chapter 4, it was discussed how the various elements of the decision tree should be analyzed in terms of the sensitivity they exhibit toward the ultimate decision goal. For the given decision problem we focused on the various costs and prior probability for the geological scenario (Figure 4.17). Similarly, a sensitivity study can be performed on the connectivity probabilities and reliability probabilities. In the latter case one can also investigate how "reliable" the data needs to be in order to change the decision one way or another, regardless of the given reliability.

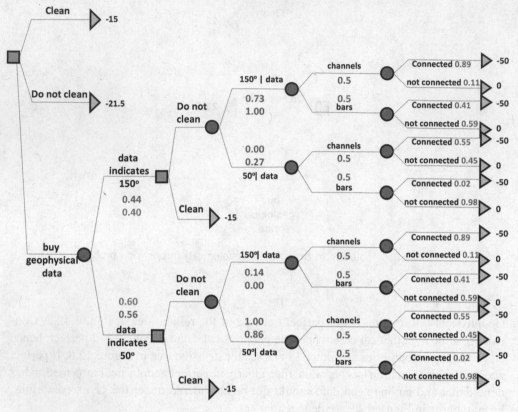

Figure 12.8   Decision tree for the value of information problem "buying geophysical data." The blue probabilities of 1 and 0 are for the case of perfect geophysical data.

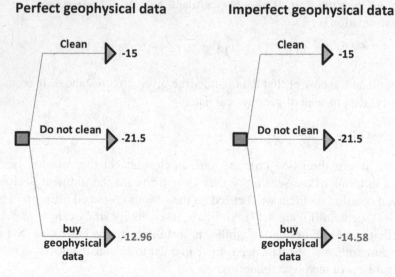

Figure 12.9   Solution of the value of (geophysical) information problem.

Note that a sensitivity study can and should be done *prior* to any major 3D Earth modeling. Such analysis will bring focus to that part of the data gathering, interpretation and Earth modeling that are most consequential to the decision making process. In fact, the relative importance of the various subjectively specified probabilities, values and costs in such analysis is often more critical than the actual absolute values entering or emanating from the decision tree.

Consider first the connectivity probabilities for which the base case is given in Figure 4.16. We study the sensitivity of these probabilities on the base decision case, that is, Figure 4.15. A change in the "total connectivity" of the system is studied by multiplying these probabilities by a certain factor a (making sure that probabilities do not exceed 1.0). Figure 12.10 shows how the cost of "do not clean" changes as function of this multiplier. In addition, the effect of connectivity is studied for each scenario independently, that is, the multiplier is applied to the first two probabilities in Figure 12.5 while the others are kept constant and vice versa. This allows studying whether connectivity for one scenario is more influential on the decision than connectivity for the other. The results in Figure 12.10 clearly shows that the channel connectivity has greater sensitivity (steeper slope) than connectivity in the bar scenario. The modeling effort (and possibly data gathering effort) should therefore be focused on the channel systems.

Next, the impact of changing the reliability probabilities is studied. For the base case, the reliabilities are symmetric (Figure 12.3), that is, the geophysical data resolves either paleo-direction equally well. Consider first keeping this symmetry by changing both

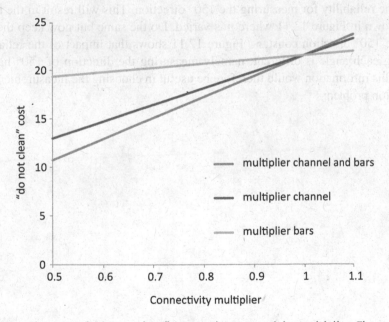

Figure 12.10 The cost of "do not clean" versus the connectivity multiplier. The multiplier is applied for three scenarios:(multiplier channel and bars) means that the multiplier is applied to the connectivity probability of both bar and channels while (multiplier channel) means it is only applied to channel connectivity probabilities.

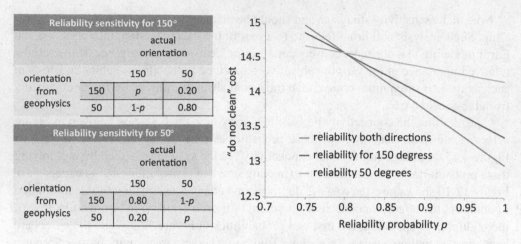

Reliability sensitivity for 150°

		actual orientation	
		150	50
orientation from geophysics	150	p	0.20
	50	1-p	0.80

Reliability sensitivity for 50°

		actual orientation	
		150	50
orientation from geophysics	150	0.80	1-p
	50	0.20	p

— reliability both directions
— reliability for 150 degress
— reliability 50 degrees

Figure 12.11 Change in the cost of "do not clean" versus a change in reliability probability.

diagonal values of 0.8 in Figure 12.3 (adjusting the off-diagonal values appropriately). Figure 12.11 shows the change in the cost of "do not clean." Consider now the case where asymmetry is introduced in the reliability probabilities, that is, the geophysical data measures each paleo-direction with different accuracy. It is, in fact, extremely common that measurement devices do not measure with equal accuracy over the measurement value range. Firstly, keep the reliability probability for the measurement 50° constant, while varying the reliability for measuring the 150° direction. This will result in the reliability matrix shown in Figure 12.11 where p is varied. Do the same but now keep the reliability for the 150° direction constant. Figure 12.11 shows that impact of the reliability for measuring each angle is different, namely measuring the direction of 150° has greater impact. This information would therefore be useful in choosing the measurement tool for this decision problem.

Index

algorithm, 114, 129–30, 184, 186
alternative Earth models, 17, 113, 134, 144, 154, 166, 181, 186
alternative models, 71, 157
aquifers, 1–3, 39, 42, 58, 70, 72, 90, 210
attributes, 59, 65–6, 135, 182–3
autocorrelation, 79, 82, 85
autocorrelation functions, 83–5
average, ensemble, 127–8
axial points, 176–7

Bayes, 19–20, 46–50, 119–23, 125–6, 128–31, 196
Bayes' Rule, 19, 118
bivariate data analysis, 33, 35
Boolean model, 89, 93–5, 108, 166
branches, 70, 74, 197–9, 220

calibration data set, 110–11, 116
Cartesian grid, 141–2, 144, 158
Cartesian space, 87, 157–8, 163, 165, 183
CCD (central composite design), 176–7, 179
cdf, empirical, 28–9, 31
climate modeling, 44, 51, 53–4, 109, 111
climate models, 8, 42, 51–3, 154, 172
cluster centers, 183–4, 186–7, 189–90
clustering, 34, 182, 185–6, 188–90
clusters, 34–5, 182–4, 186–7, 189
complexity, 2, 5, 49–50, 54, 57, 78, 146, 154
conditional probabilities, 18–19, 45, 72, 98, 109, 111–13, 116, 125–6
connectivity, 5, 134, 160, 162–3, 223
connectivity distance, 160, 162
consistency, 134, 138, 140
contaminant transport, 153, 155, 217–18
context of decision making, 55–6

continuity, 78, 113, 115
continuous random variables, 21
correlation, 33, 35, 37, 78–81, 84, 86, 103, 115
correlation coefficient, 35, 78–80, 115–16
correlation length, 80, 86
correlogram, 78, 80–2, 84, 86, 115
cost, 70, 73–6, 153–4, 197–9, 201, 216–18, 220–1, 223–4
cost/benefit, 65–6
covariance, 84–5
covariance function, 84–5

data
  hard, 7, 108, 115, 167.
  soft, 7, 115–17
data configurations, 102–3, 115
data-model relationship, 123, 166, 169
data reliability, 194–5, 201
data sources, 5–6, 46, 94, 107–11, 113, 115–16, 195, 201–2
data-variable relationship, 123, 125
decision, 3–6, 43–4, 54–62, 65–6, 70–4, 76–7, 153–5, 193–7
  binary, 64, 197, 207
  making, 7, 45, 54–70, 72, 74, 76
  optimal, 59, 70, 72, 74
decision analysis, 7–8, 56–7, 59, 67–8, 76, 194, 197–8, 215
decision goal, 55–6, 144, 154, 172, 193–5
decision maker, 59–60, 63, 198
decision model, 55, 67, 77, 93, 107, 133, 153, 171
decision node, 70–1, 73–4
decision problem, 60, 70, 108, 118, 154–5, 193, 195, 218
decision question, 4–5, 70, 119, 172, 181, 191, 195

Printed in the United States
By Bookmasters